Kainate Receptors

ADVANCES IN EXPERIMENTAL MEDICINE AND BIOLOGY

Recent Volumes in this Series

Volume 709
HISTAMINE IN INFLAMMATION
Robin L. Thurmond

Volume 710
RECENT ADVANCES ON MODEL HOSTS
Eleftherios Mylonakis

Volume 711
EPIGENETIC CONTRIBUTIONS IN AUTOIMMUNE DISEASE
Esteban Ballestar

Volume 712
CYSTEINE PROTEASES OF PATHOGENIC ORGANISMS
Mark W. Robinson and John P. Dalton

Volume 713
CELL FUSION IN HEALTH AND DISEASE, I: CELL FUSION IN HEALTH
Thomas Dittmar

Volume 714
CELL FUSION IN HEALTH AND DISEASE, II: CELL FUSION IN DISEASE
Thomas Dittmar

Volume 715
BACTERIAL ADHESION: BIOLOGY, CHEMISTRY AND PHYSICS
Dirk Linke and Adrian Goldman

Volume 716
MAST CELL BIOLOGY: CONTEMPORARY AND EMERGING TOPICS
Alasdair M. Gilfillan and Dean D. Metcalfe

Volume 717
KAINATE RECEPTORS: NOVEL SIGNALING INSIGHTS
Antonio Rodríguez-Moreno and Talvinder S. Sihra

Kainate Receptors

Novel Signaling Insights

Edited by

Antonio Rodríguez-Moreno, PhD
Departamento de Fisiología, Anatomía y Biología Celular, Universidad Pablo de Olavide, Sevilla, Spain

Talvinder S. Sihra, PhD
Department of Neuroscience, Physiology and Pharmacology, University College London, London, UK

Springer Science+Business Media, LLC

Landes Bioscience

Springer Science+Business Media, LLC
Landes Bioscience

Springer Science+Business Media, LLC, 233 Spring Street, New York, New York 10013, USA
http://www.springer.com

Please address all inquiries to the publishers:
Landes Bioscience, 1806 Rio Grande, Austin, Texas 78701, USA
Phone: 512/ 637 6050; FAX: 512/ 637 6079
http://www.landesbioscience.com

The chapters in this book are available in the Madame Curie Bioscience Database.
http://www.landesbioscience.com/curie

Kainate Receptors: Novel Signaling Insights, edited by Antonio Rodríguez-Moreno and Talvinder S. Sihra. Landes Bioscience / Springer Science+Business Media, LLC dual imprint / Springer series: Advances in Experimental Medicine and Biology.

ISBN: 978-1-4419-9556-8

While the authors, editors and publisher believe that drug selection and dosage and the specifications and usage of equipment and devices, as set forth in this book, are in accord with current recommendations and practice at the time of publication, they make no warranty, expressed or implied, with respect to material described in this book. In view of the ongoing research, equipment development, changes in governmental regulations and the rapid accumulation of information relating to the biomedical sciences, the reader is urged to carefully review and evaluate the information provided herein.

Library of Congress Cataloging-in-Publication Data

Kainate receptors : novel signaling insights / edited by Antonio Rodriguez-Moreno, Talvinder S. Sihra.
p. ; cm. -- (Advances in experimental medicine and biology ; v. 717)
Includes bibliographical references and index.
ISBN 978-1-4419-9556-8 (hardback)
1. Kainic acid--Physiological effect. 2. Kainic acid--Receptors. 3. Cellular signal transduction. I. Rodriguez-Moreno, Antonio, 1968- II. Sihra, Talvinder S., 1960- III. Series: Advances in experimental medicine and biology ; v. 717. 0065-2598
[DNLM: 1. Receptors, Kainic Acid--physiology. 2. Cerebrum--physiology. 3. Signal Transduction--physiology. W1 AD559 v.717 2011 / QV 253]
QP563.K33K35 2011
612'.015756--dc22

2011004823

DEDICATIONS

For Sabina
—Antonio Rodríguez-Moreno

For my girls, Jasmina, Anya and Ljubica
—Talvinder S. Sihra

PREFACE

This volume critically examines the functional actions of the kainate-type glutamate receptors (KARs). Following on from the larger body of work on the NMDA- and AMPA-type ionotropic glutamate receptors (GluRs), studies with KARs have consistently thrown up exceptions to general rules about synaptic modulation. Contributors herein provide an insight to the idiosyncracies that now almost typify the KAR field.

In the first chapter, Sihra and Rodriguez-Moreno set the scene with the introduction of KARs and review of the role of KAR in the modulation of interneuron GABA release in the hippocampus, highlighting some seminal and provocative early studies alluding to the metabotropic function of KARs. Continuing apace with adventurous thinking, Cherubini and colleagues consider the consequences of the inhibitory and metabotropic function of KARs in the immature hippocampus. It is now appreciated that in the developing brain, the prototypic inhibitory neurotransmitter GABA plays an incongruous excitatory role through $GABA_A$ receptors that are depolarising because of a Cl^- electrochemical equilibrium that is the reverse of that found in the adult. Remarkably in this scenario, it is glutamate, the defined "excitatory" neurotransmitter that plays the calming role through the operation of KARs that suppress the release of GABA presynaptically.

Continuing on this appraisal of the presynaptic effects of KARs, in Chapter 3, Jin and Smith show that the metabotropic functions of KARs are not restricted to hippocampal GABA release. They assess the operation of the KARs in several basal ganglia nuclei, argue for the role of direct and indirect G-protein mediation of KAR activity and highlight the subtlety of the metabotropic operation of KARs, this opening up potential pharmacotherapeutic avenues.

While metabotropic effects of KARs on GABA release provide examples of heteroreceptor regulation of neurotransmitter release, the evidence for KARs acting as autoreceptors (also in a metabotropic guise), in the regulation of glutamate release, has also accumulated. In Chapter 4, Rodriguez-Moreno and Sihra review this work which has revealed a fascinating characteristic of KARs, which is that, in some instances, they can have a bimodal action; inhibiting glutamate release in one context, while facilitating it in another, both types of modulation occurring at the same synapse.

KARs have long been remarked upon by the frank epilepsy produced by the defining agonist. Together with the aforementioned presynaptic actions of KARs, Melyan and Wheal have long suggested a postsynaptic locus of KAR action which invokes hyperexcitability. In Chapter 5, the authors provide new insights into how KARs inhibit the slow after-hyperpolarisation current and thereby control the synaptic excitability of CA1 pyramidal cells in the hippocampus. Remarkably, this modulation of postsynaptic activity by KARs is $G_{i/o}$ and PKC dependent, again alluding to the metabotropic function of the target KARs.

The molecular subunit composition of the KARs that support their metabotropic modus operandi remains enigmatic. In Chapter 6, Ruiz analyses the involvement of KARs in the synaptic transmission between dentate granule cells and CA3 pyramidal neurons, where, post-synaptically, KARs display integrative properties through a canonical ionotropic operation, but at the same time also enhance neuronal excitability through the suppression of the slow Ca^{2+} activated K^+ current I_sAHP shown to be underpinned by metabotropic characteristics. The approach taken by Ruiz and colleagues has been to dissect the function of the receptor by the analysis of effect in the hippocampus of ablating the high-affinity subunits constituting KARs. The evidence has opened up new avenues of thought in elucidating how KARs might be prone to behave like their de facto mGluRs cousins also involved in increasing neuronal excitability by suppressing K^+ conductance.

Although studies of KARs have largely concentrated on central nervous system function, clearly glutamatergic transmission also plays an essential role in peripheral nervous system and KARs are evident in the spinal cord. Accordingly, in Chapter 7, Rozas characterizes KARs in dorsal root ganglia which provide glutamatergic inputs to dorsal horn neurons. The importance of KARs in pain transmission is highlighted, not only suggesting their operation through G-protein-linked, intracellular Ca-sensitive and PKC-dependent signalling, but also the utility of the metabotropic signalling in the control of KAR receptor plasticity.

As indicated earlier, the expanding repertoire of KAR function is not limited to the adult nervous system. Indeed, the seemingly omnipotent KAR appears to be involved in the early modelling of synapses based on activity. Thus in Chapter 8, Lauri and Taira discuss the postnatal developmental role of KARs in shaping network activity. Adding another layer of modulation, the indications are that the patterns of KAR-mediated regulation may be developmentally regulated as synapses mature, so that the adult picture is very different to that prevalent postnatally.

In the penultimate Chapter 9, Plested provides a most eloquent rendition of the view of the glutamate receptor field from the structural viewpoint, but looking to use the very 21st century approach developed to elucidate modulatory mechanisms. In so doing, Plested and colleagues have opened up new horizons by their didactic consideration of how the very ions that perpetually bath neurons and are indeed conducted by ion channels like KARs play a crucial role in the maintenance and modulation of the channel activity. This sort of analysis will no doubt prove an essential strategy in the future to determine the rather unique bifurcation of the observed KAR activity modes reviewed in this volume.

Finally, whether developmentally or in the adult situation, the expression, maintenance and turnover KARs must necessarily underpin some of the long-term modulatory events associated with KAR function. In Chapter 10 Marshall, Blair and Singer provide intriguing insights into KAR interaction with intracellular binding partners, this not only determining the stability of KARs, but perhaps also forming the instrument of the physical and functional compartmentalization of the receptors in the neuron.

The fascinating insights provided by the contributors to this volume serve to encourage searching mechanistic questions. The foremost of these, arising from non-canonical (metabotropic) signalling by KARs is: what is the physical basis of KARs, having an ionotropic receptor topology, coupling to G-proteins to mediate responses independent of ionotropic activity?

The intracellular domains, present in prototypic heptahelical metabotropic receptors (G-protein coupled receptors, GPCRs), and that mediate the defining interaction with heterotrimeric G-proteins,[1] are not found in KAR subunits. Notwithstanding, unconventional G-protein interaction cannot be ruled out given the available evidence. Thus, goldfish KA-binding proteins have been reported to display PTX-sensitive agonist binding and agonist-dependent ADP ribosylation of a 40 kDa protein,[2-4] and heterologously expressed frog KARs directly bind to GTP-binding proteins.[5] One might speculate that the subunits forming the heterotetrameric KARs may contribute multiple cytoplasmic domains (including the little characterised intracellular loops of the KAR subunits) to a cryptic secondary/tertiary structure capable of binding G-proteins. Additionally, alternative splicing of Glu5/6/7 KAR subunits to produce different C-termini[6] may impart differential functional properties to the oligomeric KAR, some of which may relate to the metabotropic function of KARs.

Even if any direct interaction between KAR and G-proteins is unfeasible, indirect docking of G-proteins to KARs through intermediary adaptor proteins remains a tenable proposition.[7] Biochemical analyses may not distinguish between the two possibilities. Nonetheless, interestingly in CA3 neurons, anti-GluR6 antibody coimmunoprecipitates GluR6 subunits, KA2 subunits and $G\alpha\theta$, and thus suggest a physical association of KAR and G-protein which is contingent on the presence of the KA2 subunit.[8] Notwithstanding the more recent studies ascribing the presence of KA2 subunits to the support of ionotropic rather than metabotropic functions of KARs,[9] the initial evidence indicating the association of a G-protein with a KAR encourages the search for similar interactions of KARs with other G-proteins. In particular, a number of functional studies showing that KA-mediated modulation is pertussis-toxin-sensitive, predict some form of $G_{i/o}$ interaction with KARs, be this direct or indirect. Over the next few years, intense activity in the KAR field anticipates answers to this and many of the questions thrown up by the contributions herein. We look forward eagerly to this elucidation.

Antonio Rodríguez-Moreno, PhD
Talvinder S. Sihra, PhD

REFERENCES

1. Pin JP, Duvoisin R. The metabotropic glutamate receptors: structure and function. Neuropharmacology 1995; 34:1-26.
2. Willard JM, Ziegra, CJ, Oswald RE. The interaction of a kainate receptor from goldfish brain with a pertussis toxin-sensitive GTP-binding protein. J Biol Chem 1991; 266:10196-10200.
3. Willard JM, Ziegra, CJ, Oswald RE. The interaction of a kainate receptor from goldfish brain with a pertussis toxin-sensitive GTP binding protein. Ann N Y Acad Sci 1992; 648:351-352.
4. Ziegra CJ, Willard, JM, Oswald RE. Coupling of a purified golfish brain kainate receptor with a pertusis toxin-sensitive G protein. Proc Natl Acad Sci USA 1992; 89:4134-4138.
5. Willard JM, Oswald RE. Interaction of the frog brain kainate receptor expressed in Chinese hamster ovary cells with a GTP-binding protein. J Biol Chem 1992; 267:19112-19116.
6. Pinheiro P, Mulle C. Kainate receptors. Cell Tissue Res. 2006; 326(2):457-482.
7. Rodríguez-Moreno A, Lerma J. Kainate receptor modulation of GABA release involves a metabotropic function. Neuron 1998; 20:1211-1218.
8. Ruiz A, Sachidhanandam S, Utvik JK et al. Distinct subunits in heteromeric kainate receptors mediate ionotropic and metabotropic function at hippocampal mossy fiber synapses. J Neurosci 2005; 25:11710-11718.
9. Fernandes HB, Catches JS, Petralia RS et al. High-affinity kainate receptor subunits are necessary for ionotropic but not metabotropic signaling.

ABOUT THE EDITORS...

ANTONIO RODRÍGUEZ-MORENO is an Associate Professor and Head of the Cellular Neuroscience and Plasticity Laboratory at the University Pablo de Olavide, in Seville, Spain. His main research interests include the study of the mechanisms of synaptic plasticity in cortex and hippocampus and the physiology of glutamate receptors of the NMDA and kainate type. He is a member of the Federation of European Neuroscience Societies (FENS) and the Society for Neuroscience (SFN). Following a degree in Biological Sciences at the University of Seville, Dr. Rodríguez-Moreno received his PhD degree in Neurobiology and Molecular Biology from Cajal Institute and the University Automous of Madrid in Madrid, Spain. During a first postdoc at the University Pablo de Olavide, in Seville, Spain he studied the mechanisms of learning and memory in awake animals. This was followed by a postdoctoral stay at University College London, UK where he investigated the actions of kainate receptors using synaptosomes and brain slices. During a third postdoc at University of Oxford, he started to study the mechanisms of brain plasticity processes.

ABOUT THE EDITORS...

TALVINDER S. SIHRA is a Reader in Cellular and Molecular Neuroscience, and Teaching Lead in the Department of Neuroscience, Physiology and Pharmacology at University College London (UCL), UK. His main research interests are in the elucidation and characterisation of presynaptic receptors and downstream signalling pathways involved in the modulation of neurotransmitter release. Pursuant to an undergraduate degree in Biochemistry and Physiology from the University of Sheffield, UK, he completed his PhD (Biochemistry) in 1985 at the University of Dundee, Scotland, where he initiated his studies looking at the mechanisms underlying amino acid neurotransmitter release. He extended these studies at the Rockefeller University, New York, USA, in the laboratory of Professor Paul Greengard, Nobel Laureate in Medicine (2000), looking at the role of protein phosphorylation in presynaptic function and plasticity. In 1993, he obtained a faculty position at the University of London (Royal Free Hospital School of Medicine), and continued thereafter at UCL. Dr. Sihra is a member of the Biochemical Society (UK) and currently on the Signalling theme panel for the Society. He is a member of the Society of Neuroscience and former editor of the *British Journal of Pharmacology*.

PARTICIPANTS

Leslie A.C. Blair
Department of Molecular Pharmacology, Physiology and Biotechnology
Brown University
Providence, Rhode Island
USA

Maddalena D. Caiati
Neurobiology Sector and IIT Unit
International School of Advanced Studies (SISSA)
Basovizza (Trieste)
Italy

Enrico Cherubini
Neurobiology Sector and IIT Unit
International School of Advanced Studies (SISSA)
Basovizza (Trieste)
Italy

Xiao-Tao Jin
Division of Neuroscience
Yerkes National Primate Research Center
and
Department of Neurology
Emory University
Atlanta, Georgia
USA

Sari E. Lauri
Neuroscience Center and Department of Bio- and Environmental Sciences
University of Helsinki
Helsinki
Finland

John Marshall
Department of Molecular Pharmacology, Physiology and Biotechnology
Brown University
Providence, Rhode Island
USA

Zara Melyan
Circadian and Visual Neuroscience Group
Nuffield Laboratory of Ophthalmology
University of Oxford
The John Radcliffe Hospital
Oxford
UK

Andrew J.R. Plested
Leibniz- Institut für Molekulare Pharmakologie (FMP)
Berlin
Germany

Antonio Rodríguez-Moreno
Departamento de Fisiología, Anatomía y Biología Celular
Universidad Pablo de Olavide
Sevilla
Spain

Jose Luis Rozas
Instituto de Biomedicina de Sevilla (IBIS)
Hospital Universitario Virgen del Rocío/ CSIC/Universidad de Sevilla
and
Departmento Fisiología Médica y Biofísica
Universidad de Sevilla
and
Centro de Investigación Biomédica en Red sobre Enfermedades Neurodegenerativas (CIBERNED)
Sevilla
Spain

Arnaud Ruiz
Department of Pharmacology
The School of Pharmacy
University of London
London
UK

Talvinder S. Sihra
Department of Neuroscience, Physiology and Pharmacology
University College London
London
UK

Jeffrey D. Singer
Department of Biology
Portland State University
Portland, Oregon
USA

Sudhir Sivakumaran
Neurobiology Sector and IIT Unit
International School of Advanced Studies (SISSA)
Basovizza (Trieste)
Italy

Yoland Smith
Division of Neuroscience
Yerkes National Primate Research Center
and
Department of Neurology
Emory University
Atlanta, Georgia
USA

Tomi Taira
Neuroscience Center and Department of Bio- and Environmental Sciences
University of Helsinki
Helsinki
Finland

Howard V. Wheal
School of Biological Sciences
University of Southampton
Bassett Crescent East
Southampton
UK

CONTENTS

1. METABOTROPIC ACTIONS OF KAINATE RECEPTORS IN THE CONTROL OF GABA RELEASE 1

Talvinder S. Sihra and Antonio Rodríguez-Moreno

Abstract 1
Introduction 1
Presynaptic Modulation of GABA Release by Metabotropic Actions of KARs 2
Conclusion 8

2. IN THE DEVELOPING HIPPOCAMPUS KAINATE RECEPTORS CONTROL THE RELEASE OF GABA FROM MOSSY FIBER TERMINALS VIA A METABOTROPIC TYPE OF ACTION 11

Enrico Cherubini, Maddalena D. Caiati and Sudhir Sivakumaran

Abstract 11
Introduction 12
Presynaptic Kainate Receptors Control the Release of Glutamate and GABA 12
Kainate Receptors Modulate Cell Excitability via a Metabotropic Type of Action 13
Kainate Receptors in the Immature Hippocampus 14
GABAergic Phenotype of Immature MF Terminals 15
GluK1 KARs Down Regulate GABA Release by from Immature MF Terminals 17
Modulation of Spike-Time Dependent Plasticity by GluK1 KARs 21
Conclusion 22

3. LOCALIZATION AND FUNCTIONS OF KAINATE RECEPTORS IN THE BASAL GANGLIA 27

Xiao-Tao Jin and Yoland Smith

Abstract 27
Introduction 27

Striatum 28
Globus Pallidus 29
Substantia Nigra Pars Compacta 33
Subthalamic Nucleus 36
Conclusion 36

4. METABOTROPIC ACTIONS OF KAINATE RECEPTORS IN THE CONTROL OF GLUTAMATE RELEASE IN THE HIPPOCAMPUS 39

Antonio Rodríguez-Moreno and Talvinder S. Sihra

Abstract 39
Introduction 39
KARs on Schaffer Collateral (SC) Terminals Suppress Glutamate Release at Schaffer Collateral (SC)-CA1 Pyramidal Cell (PC) 40
KARs on Granule Cells (Dentate Gyrus) Mossy Fibre (MF) Terminals Effect Bimodal Modulation of Glutamate Release at Mossy Fibre (MF)-CA3 Pyramidal Cell (PC) Synapses 42
Conclusion 46

5. METABOTROPIC ACTIONS OF KAINATE RECEPTORS IN THE REGULATION OF I_{SAHP} AND EXCITABILITY IN CA1 PYRAMIDAL CELLS 49

Zara Melyan and Howard V. Wheal

Abstract 49
Introduction 49
Direct Postsynaptic Activation of KAR Inhibits I_{sAHP} on CA1 Pyramidal Cell 50
KARs Responsible for I_{sAHP} Inhibition Involve Metabotropic Action 53
Signal Transduction Mechanisms for Metabotropic KAR-Induced Inhibition of I_{sAHP} in CA1 Pyramidal Cells 54
Conclusion 57

6. KAINATE RECEPTORS WITH A METABOTROPIC SIGNATURE ENHANCE HIPPOCAMPAL EXCITABILITY BY REGULATING THE SLOW AFTER-HYPERPOLARIZATION IN CA3 PYRAMIDAL NEURONS 59

Arnaud Ruiz

Abstract 59
Introduction 59
High-Affinity Postsynaptic KAR Subunits in the Hippocampus 60
Action of KAR Agonists on the Slow After-Hyperpolarization in Hippocampal Pyramidal Neurons 61

Synaptically Released Glutamate Modulates I_{sAHP} in Hippocampal Pyramidal Neurons ... 62
Candidate Mechanisms Linking KAR Activation and Enhanced Cellular Excitability ... 63
Conclusion ... 63

7. METABOTROPIC ACTIONS OF KAINATE RECEPTORS IN DORSAL ROOT GANGLION CELLS ... 69

Jose Luis Rozas

Abstract ... 69
Introduction ... 69
Expression and Subunit Composition ... 70
Kainate Receptors Physiology and Plasticity at Dorsal Root Ganglion Neurons ... 72
Kainate Receptors and Pain ... 75
Conclusion ... 76

8. ROLE OF KAINATE RECEPTORS IN NETWORK ACTIVITY DURING DEVELOPMENT ... 81

Sari E. Lauri and Tomi Taira

Abstract ... 81
Introduction ... 81
Kainate Receptors and Network Activity in the Mature Hippocampus ... 82
Kainate Receptors in the Developing Brain ... 83
KAR Mediated Regulation of Network Activity in the Immature Hippocampus ... 84
Role of KARs in Developmental Maturation of Synaptic Connectivity ... 86
Conclusion ... 88

9. KAINATE RECEPTOR MODULATION BY SODIUM AND CHLORIDE ... 93

Andrew J.R. Plested

Abstract ... 93
Introduction ... 93
Physical Characteristics of Sodium and Chloride ... 93
Structural View of Glutamate Receptor Activation ... 94
Ion Binding Sites in the Dimer Interface of Kainate Receptors ... 96
The Role of Ions in Kainate Receptor Gating ... 99
Allosteric Modulation of Other Ion Channel Receptors by Monovalent Ions ... 104
Physiological Situation with Regard to Ions ... 106
Conclusion ... 109

10. BTB-KELCH PROTEINS AND UBIQUITINATION OF KAINATE RECEPTORS 115

John Marshall, Leslie A.C. Blair and Jeffrey D. Singer

Abstract........ 115
Introduction........ 115
KAR Regulation by the Ubiquitin-Proteasome Pathway........ 117
AF Regulation of Shank, a Parallel Path to Regulate Surface GluR Expression........ 119
Development of Novel Peptidomimetic Drugs for the Treatment of Neurological Disorders........ 119
Conclusion........ 121

11. CONCLUDING REMARKS........ 127

Antonio Rodríguez-Moreno and Talvinder S. Sihra

INDEX........ 129

CHAPTER 1

Metabotropic Actions of Kainate Receptors in the Control of GABA Release

Talvinder S. Sihra* and Antonio Rodríguez-Moreno*

Abstract

Kainate receptors (KARs) are members of the family of ionotropic glutamate receptors (iGluRs) which also include NMDA and AMPA receptors. As ionotropic receptors, KARs have been characterized, pre and postsynaptically, in several brain regions. In this chapter we review evidence that suggests that KARs mediate some of their effects without invoking ion-fluxes. Beginning with seminal experiments described some ten years ago, when the notion of a metabotropic action of KAR was first posited in the modulation of GABA release from hippocampal interneurons, increasingly, there have been reports indicating that some KAR functions overtly depend on G-protein activation and involve the participation of intracellular signalling cascades. Thus, KAR activation instigates a cascade involving $G_{i/o}$, phospholipase C and protein kinase C to suppress the release of GABA and therefore underpins disinhibition of pyramidal cells in the CA1 region of the hippocampus. This type of metabotropic function of KARs in controlling GABA release represents an additional level of activity-dependent control of synaptic inhibition which is independent of any ionotropic activity of KARs.

Introduction

Gamma-aminobutyric acid (GABA) is the major inhibitory neurotransmitter in the mammalian central nervous system. Excitatory (glutamatergic) inputs onto the cell bodies of interneurons instigate stimulation of these neurons, whereupon, GABA release is evoked at nerve terminals. Released GABA elicits inhibitory postsynaptic potentials (sIPSPs) in the recipient neuron through ionotropic $GABA_A$ receptor activation. The control of GABA release therefore represents a key locus for the control of synaptic network activity. In order to investigate factors underpinning the modulation of GABA release, synaptic circuits which culminate in the generation of IPSPs are particularly useful to give a quantifiable electrophysiological read-out of presynaptic activity in interneurons. In the hippocampus, interneurons in the *stratum oriens* project onto CA1 pyramidal cells (CA1 PC), to produce IPSPs when GABA is released (Fig. 1, inset). Similarly, interneurons in the *stratum radiatum* also synapse onto CA1 PCs to produce inhibition (Fig. 1, inset). The upstream initiation of interneuron activity is logically through the activation of glutamatergic circuits leading to the glutamate release and stimulation of excitatory glutamate receptors. The

*Corresponding Authors: Antonio Rodríguez-Moreno—Departamento de Fisiología, Anatomía y Biología Celular, Universidad Pablo de Olavide, Ctra. de Utrera, Km. 1, 41013, Sevilla Spain. Email: arodmor@upo.es
Talvinder S. Sihra— Department of Neuroscience, Physiology and Pharmacology, University College London,Gower Street, London, UK. Email: t.sihra@ucl.ac.uk

Kainate Receptors: Novel Signaling Insights, edited by Antonio Rodríguez-Moreno and Talvinder S. Sihra.

question arises, is the primary excitatory *ionotropic* activity of glutamate on interneuron cell bodies the only glutamatergic activity? A number of studies now point to the possibility that, together with the fast ionotropic activation of interneurons, glutamate release also regulates interneuron activity manifest in GABA release, through heteroreceptors with modulatory influences displaying slow time-courses of activation and deactivation.

The two major families of glutamate receptors consist of three ionotropic, viz. NMDA-, AMPA- and Kainate-type, glutamate receptors (iGluRs) and three sub-groups, viz. Types I, II and III,[1,2] of seven-transmembrane region, G-protein-coupled metabotropic glutamate receptors (mGluRs). The iGluRs are hetero- or homo-meric tetramers, composed of multiple, often alternatively spliced and edited subunits (NMDA receptors—NR1,2A-D: AMPA receptors—GluR1-4: Kainate receptors (KARs)—GluR-7, KA1-2). All the iGluRs are found distributed pre and postsynaptically to varying degrees. The ionotropic roles of NMDA and AMPA receptors are well recognized in supporting synaptic transmission and the control of neuronal excitability. KARs, on the other hand, have remained more incongruous, not least because of their low single channel conductance, counterpointed by a somewhat extended time-course of activity.[3,4] Development of pharmacological tools over the last 10-12 years, particularly the AMPA receptor antagonists GYKI53655 and SYM2206, have aided differentiation of KAR and AMPA receptors, which otherwise display cross-activation by their defining agonists, thus confounding definitive analysis. Additionally now, transgenic mice with specific KAR subunits ablated have greatly accelerated the elucidation of the role of KARs in central nervous system (CNS) physiology.

KARs are ubiquitously distributed in the CNS and classically shown to be expressed postsynaptically in the principal cells and interneurons of the hippocampus, lateral amygdala, dorsal root ganglia, bipolar cells of the retina, cerebral cortex, globus pallidus and cerebellum.[5] An important outcome of recent studies has been that there is now a definitive case for the presynaptic actions of KARs at several synapses. Accumulating data indicate that KARs are significantly expressed at presynaptic terminals and underpin significant modulation of neurotransmitter release therein.[4,5]

Pre and postsynaptically, while the actions of KARs have classically been attributed to the canonical ionotropic activity associated with typical iGluRs, significantly, the effects of KAR activation have often been found to be temporally inconsistent with ionotropic mechanisms operating within millisecond time-scales. Indeed frank sensitivity to agents affecting heterotrimeric G-proteins function invokes overt metabotropic actions of KARs, without the receptors themselves being GPCRs per se (i.e., single proteins with seven-transmembrane regions and defined intracellular sites for G-protein interaction). While this notion may be somewhat unusual, significantly, the precedence already exists from studies with sister AMPA receptors which also evidently mediate some of their effects by metabotropic operation.[6-13] We review here the evidence for KARs modulating the release of GABA in their metabotropic guise, while independently supporting canonical ionotropic functions.

Presynaptic Modulation of GABA Release by Metabotropic Actions of KARs

Initial evidence of a metabotropic function of KAR in the brain was delineated by studies looking at hippocampal synapses established between interneurons, with interneuron cell bodies in *stratum oriens* or *stratum radiatum*, projecting to CA1 pyramidal cells (CA1 PCs)[14-16] (Fig. 1, inset). GABA release at these synapses is assessed by measuring the evoked inhibitory postsynaptic currents (eIPSC) in the CA1 PCs consequent from inhibitory $GABA_A$ receptor activation. Application of kainate (KA) results in a depression of the eIPSC which is reversed on the washout of the drug (Fig. 1A). This presynaptic modulation of GABA release by KARs, displays several interesting properties that allude to a metabotropic mechanism of KAR coupling to the decrease in GABA release. Firstly, the inhibition of synaptic depression by KAR activation is abolished by pertussis toxin (PTX), indicating the involvement of a pertussis toxin (PTX)-senstive G-protein such as G_i or G_o (Fig. 1B,C). Secondly, the regulation by KAR activation is severely attenuated by inhibitors of phospholipase C (PLC) and the protein kinase C (PKC) inhibitors, staurosporine, calphostin C or bisindolylmaleimide (Fig. 1C).

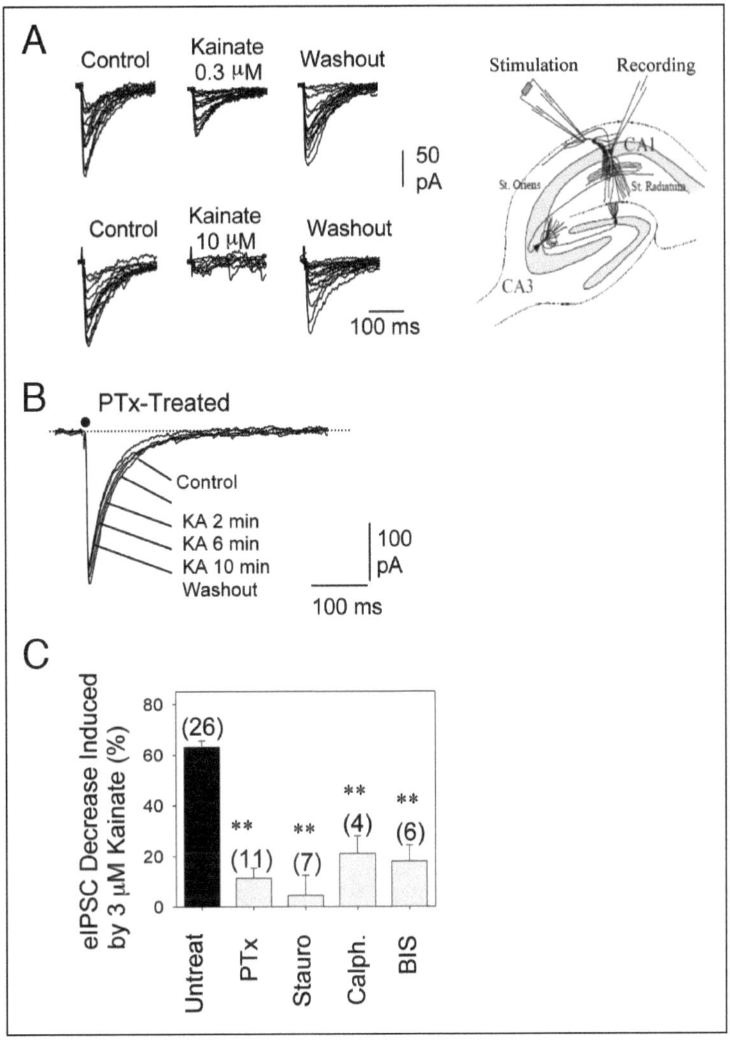

Figure 1. Metabotropic Kainate receptor-mediated depression of GABA release in the CA1 region of the hippocampus. Inset: Experimental configuration. A) Evoked IPSC (eIPSC) records showing KAR activation reversibly depresses GABAergic transmission at the interneuron-CA1 hippocampal synapse in slices in the presence of AMPA and NMDA receptors antagonists. KA at 10 μM produces an almost complete reduction of the eIPSCs amplitude. B) The action of KA involves the activation of a $G_{i/o}$-protein. In slices incubated with pertussis toxin (PTX), the depressant action of KA on the eIPSC is not observed (cf. (a)). C) In addition of the $G_{i/o}$-protein dependence, the depression of the eIPSCs amplitude mediated by KAR activation involves the activation of PKC. Thus the KAR-mediated depression is significantly reduced in the presence of the PKC inhibitors, staurosporine (Stauro, 0.5 μM), calphostin C (Calph C, 0.5 μM) and bisindolylmaleimide (BIS, 0.1 μM). Panel (A) reproduced from Rodríguez-Moreno A et al. Neuron 1997; 19:893-901,[14] ©1997 with permission from Elsevier. Panels (B) and (C) reproduced from Rodríguez-Moreno A et al. Proc Natl Acad Sci USA 2000; 97:1293-1298,[21] ©2000 with permission from the National Academy of Sciences, USA.

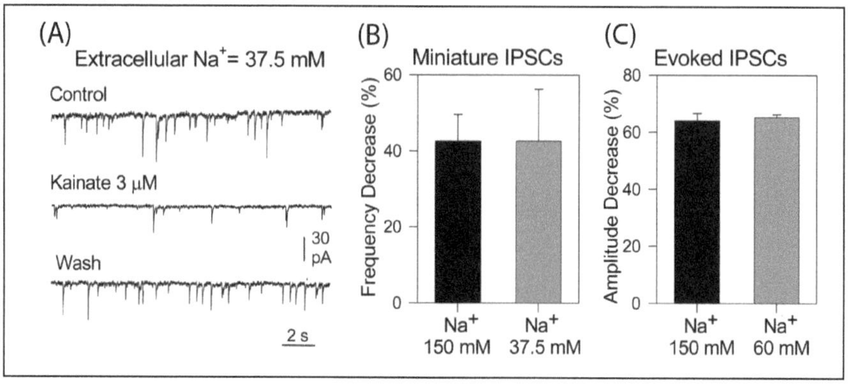

Figure 2. The downregulation of GABA release by KA is independent of ion channel activity. A) Records showing the effect of KA in the presence of low extracellular Na^+ (25% of total) intended to reduce ionotropic KAR activity. In this low [Na^+] condition, the application of KA produces a similar change in the frequency of miniature IPSCs (B) and in the evoked IPSC amplitude (C), as in normal extracellular solution. Panels (A),(B) and (C) reproduced from Rodríguez-Moreno A, Lerma J. Neuron 1998; 20:1211-1218,[16] ©1998 with permission from Elsevier.

To obviate the possibility that the observed modulatory activity is a consequence of the canonical ionotropic activity predicted of iGluRs, investigators have measured the miniature IPSC frequency (as readout of spontaneous action potential independent presynaptic activity) and eIPSC frequency, in the presence of normal or reduced extracellular [Na^+]. The rationale of the experiment is that, if KARs reduce IPSCs by ionotropic means, reduction of the conducting Na^+ ion should mitigate against the reduction of GABA release. Crucially, the lowering of extracellular [Na^+] is of no consequence on the effect of KA reducing either the miniature or the evoked IPSC (Fig. 2). This therefore points to a mechanism of KAR action which is independent of any ionotropic activity. The model proposed from these data therefore invokes that presynaptic KARs located on inhibitory terminals activate $G_{i/o}$, which, through its βγ-subunits, couples to phospholipase C (PLC) stimulation. The consequent production of the second messenger diacylglycerol (DAG) causes downstream activation of a pool of PKC that results in a decrease of GABA release by phospho-regulation of downstream effectors[16] (Fig. 4).

While the pharmacological activation KARs by exogenous KA points to the involvement of the receptors in the modulation of GABA release, direct support for the presence and function of KARs at interneuron-CA1 principal cell synapses warrants demonstration that the response produced by exogenously applied KA can be recapitulated with synaptically released glutamate. Experiments by Min et al[17] address this by measuring the eIPSC in CA1 PCs produced monosynaptically by electrically stimulated *stratum radiatum* interneurons (Fig. 3, Proximal electrode), before and after the stimulation of glutamate release elicited by a distal electrode in the Schaffer collateral axon field projecting to CA1 PCs (Fig. 3, Distal). In this experimental configuration, picrotoxin-sensitive test eIPSCs are depressed by synaptically released, endogenous glutamate (evoked by conditioning stimulation of glutamatergic afferents in CA1 region of the hippocampus; Fig. 3A) and this disinhibition is suppressed by KAR antagonism by DNQX (Fig. 3A,B; DNQX is a KAR antagonist in this context, because AMPA receptors are already blocked). Given that there is no evidence for axo-axonic synapses of glutamate terminals onto interneurons, the results with the experimental paradigm used suggest that a "spillover" of glutamate from glutamatergic terminals on CA1 PCs might heterosynaptically activate KARs on interneuron terminals to modulate GABA release (Fig. 3). Interestingly, pertaining to the subcellular mechanism of KAR action, the studies by Min et al

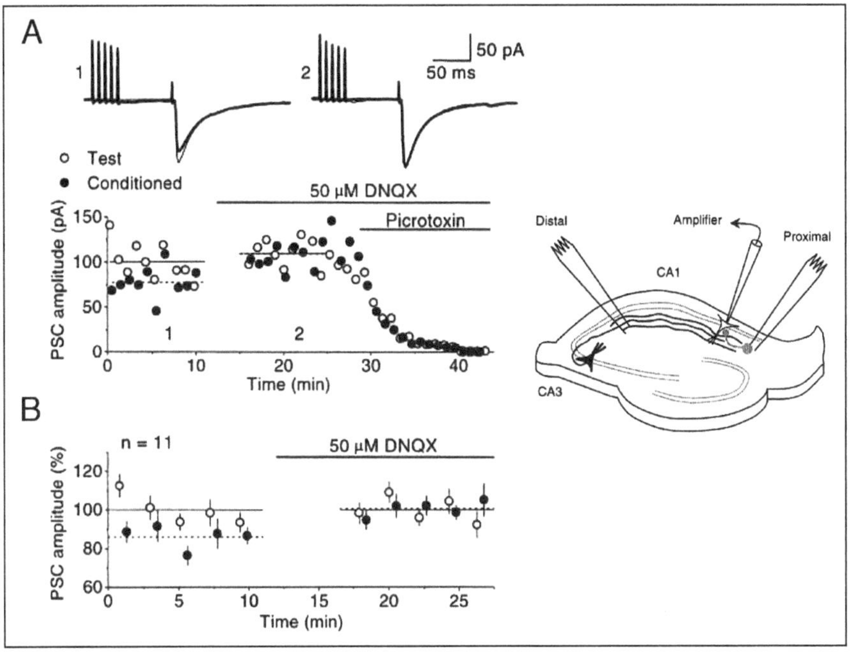

Figure 3. Depression of GABA release by endogenous glutamate activating kainate receptors. Inset: Experimental configuration. Schematic illustration of the hippocampal slice showing the positioning of the recording pipette and stimulating electrodes. Schaffer collateral axons are shown in black and a local inhibitory interneuron is shown in gray. A) Glutamate released by activating Schaffer collaterals with a train of stimuli reduces the amplitude of eIPSCs evoked by a single stimulus at *stratum oriens* (upper panel, bold trace); thus glutamate mimicks the effect of the exogenous agonist KA seen in Figure 1A. B) The decrease is present in the presence of AMPA receptor antagonists, but is sensitive to DNQX (lower panel). Reproduced from Min MY et al. Proc Natl Acad Sci USA 1999; 96:9932-9937,[17] ©1999 with permission from the National Academy of Sciences, USA.

(1999)[17] note that the depression of the IPSCs outlasts the conditioning stimulation. This is therefore suggestive of a persistent metabotropic action of KARs on interneurons, contrasting with an ionotropic action whereby any KAR-mediated currents would typically be expected to deactivate with rapid (ms) kinetics.

Although, the collective data with applied KA and synaptically released glutamate point to a reduction of GABA release by presynaptic KARs, under similar stimulation paradigms,[18] somatodendritic KARs are also activated. This leaves the question begging: are the KARs which effect a decrease in eIPSC at interneurons synapses, somatodendritic or nerve terminal-resident? Indeed, despite the pharmacological support for the inhibition of GABA release, at *stratum oriens* or *stratum radiatum* interneuron-CA1 PC synapses, being mediated by presynaptic KARs activating $G_{i/o}$ coupled to PLC and downstream PKC (Figs. 1, 4), there have been several contentious viewpoints questioning the existence of metabotropic actions of KARs, instead favouring a canonical ionotropism-based rationalization of the reported modulation of GABA release. Thus, based on the observation that ionotropic KARs do indeed occur in the somatodendritic compartment of interneurons, Frerking et al (1999)[19] propose that the activation of these KARs causes an initial increase in GABA release, whereupon there is an activation of inhibitory presynaptic, metabotropic $GABA_B$ receptors which suppresses GABA release—explaining the inhibitory

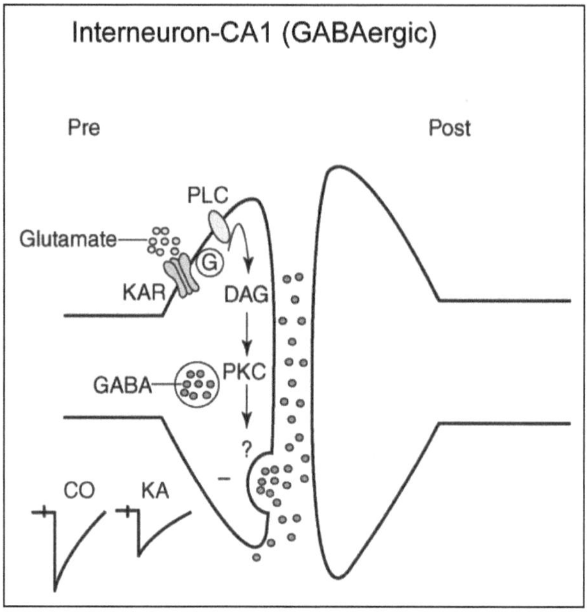

Figure 4. Metabotropic actions of KARs at *stratum oriens (or radiatum)* interneuron-CA1 synapses. Presynaptic KARs activation attenuates GABA release from interneneuron terminals (Pre) onto CA1 pyramidal cell dendrites (Post). Modulation involves a presynaptic pertussis toxin-sensitive G protein coupled to phospholipase C (PLC) which generates 2nd messenger diacylglycerol (DAG) to activate protein kinase C (PKC). PKC phosphorylates an as yet unknown target(s), to decrease GABA release. Modified and reproduced from Rodríguez-Moreno, Sihra TS. Trends Neurosci 2007; 30(12):630-637,[37] ©2007 with permission from Elsevier.

effect of KARs on the eIPSC recorded in CA1 PCs. Together with this indirect effect of KARs through an initial GABA release stimulating presynaptic $GABA_B$ receptors,[19] it is also suggested that activation of postsynaptic $GABA_A$ receptors (by the initial release of GABA) and KARs may contribute to the observed reduction in the IPSC by a respective change in series resistance and electrical shunting which then mitigate the eIPSP to produce the observed disinhibition seen with KA application.[20] Although, the foregoing scenario is consistent with the activation of KARs producing an increase in the spontaneous inhibitory postsynaptic current (sIPSC) frequency on the one hand, while decreasing the eIPSC amplitude on the other hand, there are several lines of evidence which argue against these contentions.

Firstly, a number studies show that the KAR-mediated inhibition of GABA release persists in the presence of a diverse range of $GABA_B$ receptors blockers.[14,15,17,21,22] This, therefore, obviates the proposal that the inhibitory effect of KARs on GABA release is indirectly mediated by the activation of $GABA_B$ autoreceptors by an initial burst of GABA release. An ionotropically initiated mechanism does not therefore satisfactorily explain the KA-induced reduction of the eIPSC amplitude at hippocampal interneuron-CA1 PC synapses. This is at least certainly true for the hippocampal interneuron-CA1 PC synapse, though interestingly, at spinal cord inhibitory neurons, the decrease in eIPSC amplitudes produced by KA does indeed involve a $GABA_B$ receptor-mediated feedback inhibition of GABA release.[23]

A second line of contention against an ionotropism-based mechanism for KARs in inhibiting GABA release comes from studies by Rodríguez-Moreno et al,[21] who demonstrate that the KA-elicited increase of the sIPSC frequency and the decrease of the eIPSC amplitude in *stratum oriens* interneurons, are independent phenomena which can be dissociated by the differential use of agonists. Thus application of 3-10 μM of the endogenous agonist glutamate causes a clear decrease of eIPSC amplitude, but has no effect on the sIPSC frequency. On the other hand, 0.3 μM ATPA (a GluR5 selective agonist) produces an unambiguous increase in the sIPSC frequency, but without effect on the eIPSC amplitude. This strongly suggests that the two effects of KAR activation on the sIPSC (somatodendritic) and eIPSC (nerve terminal) are independent events. Tellingly, when KA is applied in the presence of PTX, the agonist increases the sIPSC, despite the modulatory (metabotropic) effect on the eIPSC being abolished by PTX inhibition of $G_{i/o}$. This is clear verification that the ionotropic activation of KARs produces an increase in sIPSCs which is separable from the activation of terminal resident KARs. Thus in conclusion, while somatodendritic KAR can clearly operate to increase interneuron excitability, inhibition of GABA release is achieved by a different population of KARs located at the interneuron terminals, these receptors displaying a metabotropic *modus operandi*.

Supporting evidence for nerve terminal-resident KARs producing G-protein dependent inhibition of GABA release, as postulated by Rodríguez-Moreno and Lerma,[16] has come from biochemical studies with isolated nerve terminals (synaptosomes). Synaptosomes, by definition, are devoid of a somatodendritic compartment and can thus only reflect nerve terminal resident receptor activity. Using the synaptosomal preparation,[23-25] or interneuron microcultures where the presynaptic cell could be monitored,[14] KA application produces a robust decrease of endogenous GABA release. These experiments firstly demonstrate that the absence of somatodendritic KARs does not affect the inhibitory effect of KAR activation and therefore ionotropic activity is not a prerequisite for the modulation. Moreover, having demonstrated a coupling of KARs to $G_{i/o}$ activity in hippocampal membranes,[26] Cunha et al[27] further show that the KAR-mediated depression of GABA release, from hippocampal synaptosomes is PTX- and PLC inhibitor-sensitive. These data together, underline the postulate that the inhibition of GABA release at interneuron-CA1 PC synapses is likely mediated by a novel metabotropic action of KARs present in nerve endings per se, without the need for prior ionotropic activity consequent from somatodendritic KAR activation.

The separation of somatodendritic and nerve terminal KAR activities observed in the adult hippocampus,[21] had been recapitulated in the developing hippocampus. Thus with GABAergic transmission at neonatal CA1 pyramidal neurons, KAR activation elicits a very large increase in the sIPSCs frequency, but at the same time robustly depresses the eIPSC.[28] The question remains as to the developmental relevance of contrasting and opposing effects of KAR activation. One possibility is that differential activity may arise due to two types of KARs responding to different patterns of neuronal activity. For example synchronous activation of numerous CA3 inputs might induce large-scale activity/glutamate release onto CA1 neurons and thereby activate somatodendritic/axonal KARs to increase GABA release. Nerve terminal KARs on the other hand would possibly sense the local spill-over of glutamate from neighbouring glutamatergic afferents to provide a spatially restricted suppression of GABA. In this dual capacity, KARs may therefore play a crucial role in the development and maturation of hippocampal circuits, given that the actions of GABA are critical to normal network establishment and function.[28] In the adult hippocampus, this and alternative interpretations can be posited for the opposing effects of KARs on interneurons in the physiological context.[21] Notwithstanding, the pharmacological observations are in keeping with the role of KARs in the control of excitability in the adult brain.

At the molecular level, the activation of two KAR populations with opposing actions, located in separate subcellular compartments of the same interneurons, is likely underpinned by properties imparted by different subunit compositions of the receptors and/or intra/extracellular interactions that affect receptor function. Pertinent to this functional compartmentalization of KARs are the concentrations of glutamate that physiologically activate the receptors following synaptic glutamate release and subsequent "spillover" of neurotransmitter to nonsynaptic locations. From transgenic mice studies, it would appear that both GluR5- and GluR6-subunits are required to invoke interneuron-CA1 synaptic sensitivity to KA. Coassembly with one or other of these subunits with KA2 would be predicted to produce higher affinity KARs, perhaps those in the nerve terminals, which possibly respond to spillover glutamate and thus mediate a decrease of GABA release following intense glutamatergic stimulation. Unfortunately, compensatory changes in transgenic mice with GluR5 or GluR6 KAR subunits ablated confound the interpretation of data that might otherwise allow the elucidation of the subunit composition of native KARs.[29] Consequently, despite providing some novel insights, knockout studies and current knowledge of KAR localization/trafficking, cannot yet predict any contingency of the distinct agonist affinities of somatodendritic and terminal KARs, with specific subunits compositions.[29,30]

Together with the now compelling evidence for metabotropic disinhibition produced by a presynaptic KAR, there is also some evidence of KARs *increasing* GABA release from interneurons. Thus, low KA concentrations produce a consistent increase in spontaneous GABA release at CA1 interneuron-interneuron synapses.[31] Interestingly, paired recordings of interneuron-CA1 PC synapses indicate that, while the low concentrations of KA mediate facilitation of release in pairs exhibiting low initial release probability, those with high release probability are actually inhibited by KAR *antagonism*,[32] as well as being susceptible to inhibition at higher KA concentrations. The former effect of KAR antagonism may be a "defacilitation" resultant from the tonic activity of somatodendritic KARs in interneurons with high release probability, perhaps invoked by glutamate spillover from nearby synaptic terminals or from nonsynaptic sources such as astrocytes.[33] The facilitatory effects of KARs are generally consistent with ionotropic effects of extrasynaptic somatodendritic KAR. Notwithstanding, there are two important points of note arising from these studies. First, the facilitation of interneuron-interneuron synapses by KAR activation is not sensitive to Ca^{2+} channel block by Cd^{2+} (ref. 31). This is somewhat surprising if the KARs are operating ionotropically in the observed facilitation of GABA release. Second, in experiments looking at connected pairs of interneurons and CA1 PCs, the facilitatory effect of KA is relatively long-lasting.[32] Thus, even the enhancement GABA release by KARs displays features of metabotropic function. It remains to be seen in future studies whether this case develops as compellingly as the one for metabotropic KAR-mediated inhibition of GABA release.

Here, we have largely discussed specific examples of GABA release being modulated by KARs in the hippocampus, where KARs appear to function by a metabotropic *modus operandi* at interneuron-CA1 PC synapses and possibly, at interneuron-interneuron synapses. Other instances of KAR-mediated regulation of GABA release in the cortex[34] (inhibitory effect), amygdala[35] (bidirectional effect), striatum[36] (indirect inhibitory effect), spinal cord[23] (indirect inhibitory effect) and hypothalamus[33] (facilitatory effect), appear to reflect exclusively canonical ionotropic operation of KARs, or evince mechanisms where an ionotropic action underpins downstream metabotropic signalling. Examples of the latter include KARs mediating indirect metabotropic influences through, for instance, adenosine receptors in striatal cells[36] and $GABA_B$-receptors in spinal interneurons.[23] Altogether, the accumulated data point to multimodal operation of KARs by ionotropic, metabotropic and indirect-metabotropic mechanisms operating in parallel to mediate the activity-dependent tuning of synaptic inhibition by bidirectional modulation of GABA release.

Conclusion

In conclusion, KARs appear to provide a much richer spectrum of regulatory influence at GABAergic synapses than might have been predicted from their structural design. While nerve terminal resident KARs in the hippocampus evidently invoke a metabotropic influence to effect a suppression of GABA release at interneuron-CA1 PC synapses and thereby produce disinhibition,[37,38] this regulation can occur independently in the same neurons expressing somatodendritic KARs which are stimulatory. The functional expression of this opposing heterosynaptic regulation of GABA release may subserved by the activity-dependent regulation of glutamate release from terminals projecting to CA1 PCs. At CA1 interneuron-interneuron synapses in the hippocampus, the observed KAR-mediated presynaptic enhancement of GABAergic transmission would also be disinhibitory in terms of net output, but this type of modulation by KARs awaits further mechanistic elucidation.

Acknowledgements

Work performed by the authors concerning metabotropic actions of KARs was supported by Grant BFU2006-14155 from the Spanish MEC (ARM) and a FEBS Short-term Fellowship to AR-M, held in the laboratory of TSS.

References

1. Hollmann M, Heinemann S. Cloned glutamate receptors. Annu Rev Neurosci 1994; 17:81-108.
2. Jonas P, Monyer H. Ionotropic glutamate reeptors in the CNS. Berlin:Springer, 1999.
3. Lerma J, Paternain AV, Rodríguez-Moreno A. López-García JC. Physiol Rev 2001; 81:971-998.
4. Lerma J. Role and rules of kainate receptors in synaptic transmission. Nat Rev Neurosci 2003; 4:481-495.
5. Huettner JE. Kainate receptors and synaptic transmission. Prog Neurobiol 2003; 70:387-407.
6. Wang Y, Durkin JP. α-amino-3-hydroxy-5-methyl-4-isoxazolepropionic acid, but not N-mathyl-D-aspartate, activates mitogen-activated protein kinase through G-protein βγ subunits in rat cortical neurons. J Biol Chem 1995; 270:22783-22787.
7. Wang Y, Small, DL, Stanimirovic DB et al (1997) AMPA receptor-mediated regulation of a G_i-protein in cortical neurons. Nature 1997; 389:502-504.
8. Hayashi T, Umemori H, Mishina M et al. The AMPA receptor interacts with and signals through the protein tyrosine kinase Lyn. Nature 1999; 397:72-76.
9. Kawai F, Sterling P. AMPA receptor activates a G-protein that suppresses a cGMP-gated current. J Neurosci 1999; 19:2954-2959.
10. Perkinton MS, Sihra TS, Willians RJ. Ca^{2+}-permeable AMPA receptors induce phosphorilation of cAMP response element-binding protein through a phosphatidylinositol 3-kinase-dependent stimulation of the mitogen-activated protein kinase signalling cascade in neurons. J Neurosci 1999; 19:5861-5874.
11. Marin P, Fagni L, Torrens Y et al. AMPA receptor activation induces association of G-beta protein with the alpha subunit of the sodium channel in neurons. Eur J Neurosci 2001; 14:1953-1960.
12. Satake S, Saitow F, Rusakov D et al. AMPA receptor-mediated presynaptic inhibition at cerebellar GA-BA-ergic synapses: a characterization of molecular mechanisms. Eur J Neurosci 2004; 19:2464-2474.
13. Takago H, Nakamura Y, Takahashi T. G protein-dependent presynaptic inhibition mediated by AMPA receptors at the calyx of Held. Proc Natl Acad Sci USA 2005; 102:7368-7373.
14. Rodríguez-Moreno A, Herreras O, Lerma J. Kainate receptors presynaptically downregulate GABAergic inhibition in the rat hippocampus. Neuron 1997; 19:893-901.
15. Clarke VRJ, Ballyk BA, Hoo KH et al. A hippocampal GluR5 kainate receptor regulating inhibitory synaptic transmission. Nature 1997; 389:599-603.
16. Rodríguez-Moreno A, Lerma J. Kainate receptor modulation of GABA release involves a metabotropic function. Neuron 1998; 20:1211-1218.
17. Min MY, Melyan Z, Kullmann D. Synaptically released glutamate reduces γ-amonobutiric acid (GABA)ergic inhibition in the hippocampus via kainate receptors. Proc Natl Acad Sci USA 1999; 96:9932-9937.
18. Semyanov A, Kullmann DM. Kainate receptor-dependent axonal depolarization and action potential initiation in interneurons. Nat Neurosci 2001; 4:718-723.
19. Frerking M, Petersen CC, Nicoll RA. Mechanisms underlying kainate receptor-mediated disinhibition in the hippocampus. Proc Natl Acad Sci USA 1999; 96:12 917-12 922.
20. Bureau I, Bischoff S, Heinemann SF et al. Kainate receptor-mediated responses in the CA1 field of wild-type and GluR6-deficient mice. J Neurosci 1999; 9:653-663.
21. Rodríguez-Moreno A, López-García JC, Lerma J. Two populations of kainate receptors with separate signalling mechanisms in hippocampal interneurons. Proc Natl Acad Sci USA 2000; 97:1293-1298.
22. Frerking M, Malenka RC, Nicoll RA. Synaptic activation of kainate receptors on hippocampal interneurons. Nat Neurosci 1998; 1:479-486.
23. Kerchner GA, Wang GD, Qiu CS et al. Direct presynaptic regulation of GABA/glycine release by kainate receptors in the dorsal horn: an ionotropic mechanism. Neuron 2001; 32:477-488.
24. Cunha RA, Constantino MD, Ribeiro JA. Inhibition of (3H)gamma-aminobutiric acid release by kainate receptor activation in rat synaptosomes. Eur J Pharmacol 1997; 323:167-172.
25. Perkinton MS, Sihra TS. A high-affinity presynaptic kainate-type glutamate receptor facilitates glutamate exocytosis from cerebral cortex nerve terminals (synaptosomes). Neuroscience 1999; 90:1281-1292.
26. Cunha RA, Malva JO, Ribeiro JA. Kainate receptors coupled to G(i)/G(o) proteins in the rat hippocampus. Mol Pharmacol 1999; 56:429-33.

27. Cunha RA, Malva JO, Ribeiro JA. Pertussis toxin prevents inhibition by kainate receptors of rat hippocampal (3H) GABA release. FEBS Letters 2000; 469:159-162.
28. Maingret F, Lauri SE, Taira T et al. Profound regulation of neonatal CA1 rat hippocampal GABAergic transmission by functionally distinct kainate receptor populations. J Physiol 2005; 567:131-142.
29. Christensen JK, Paternain AV, Selak S et al. A mosaic of functional kainate receptors at hippocampal interneurons. J Neurosci 2004; 24:8986-8993.
30. Jaskolski F, Coussen F, Nagarajam N et al. Subcellular localization and trafficking of kainate receptors. Trends Pharmacol Sci 2005; 26:20-26.
31. Cossart R, Tyzio R, Dinocourt C et al. Presynaptic kainate receptors that enhance the release of GABA on CA1 hippocampal interneurons. Neuron 2001; 29:497-508.
32. Jiang L, Xu J, Nedergaard M et al. A kainate receptor increases the efficacy of GABergic synapses. Neuron 2001; 30:503-513.
33. Liu QS, Patrylo PR, Gao XB et al. Kainate acts at presynaptic receptors to increase GABA release from hypothalamic neurons. J Neurophysiol 1999; 82:1059-62.
34. Ali AB, Rossier J, Staiger JF et al. Kainate receptors regulate unitary IPSCs elicited in pyramidal cells by fast-spiking interneurons in the neocortex. J Neurosci 2001; 21:2992-2999.
35. Braga MF, Aroniadou-Anderjaska V, Xie J et al. Bidirectional modulation of GABA release by presynaptic glutamate receptor 5 kainate receptors in the basolateral amygdala. J Neurosci 2003; 23:442-452.
36. Chergui K, Bouron A, Normand E et al. Functional GluR6 kainate receptors in the striatum: indirect downregulation of synaptic transmission. J Neurosci 2000; 20:2175-2182.
37. Rodríguez-Moreno, Sihra TS. Kainate receptors with a metabotropic modus operandi. Trends Neurosci 2007; 30(12):630-637.
38. Rodríguez-Moreno A, Sihra TS. Metabotropic actions of kainate receptors in the CNS. J Neurochem 2007; 103:2121-2135.

Chapter 2

In the Developing Hippocampus Kainate Receptors Control the Release of GABA from Mossy Fiber Terminals via a Metabotropic Type of Action

Enrico Cherubini,* Maddalena D. Caiati and Sudhir Sivakumaran

Abstract

Kainate receptors (KARs) are glutamate-gated ion channels assembled from various combinations of GluK1-GluK5 subunits with different physiological and pharmacological properties. In the hippocampus, KARs expressed at postsynaptic sites mediate a small component of excitatory postsynaptic currents while at presynaptic sites they exert a powerful control on transmitter release at both excitatory and inhibitory connections. KARs are developmentally regulated and play a key role in several developmental processes including neuronal migration, differentiation and synapse formation. Interestingly, they can signal through a canonical ionotropic pathway but also through a noncanonical modality involving pertussis toxin-sensitive G proteins and downstream signaling molecules.

In this Chapter some of our recent data concerning the functional role of presynaptic KARs in regulation of transmitter release from immature mossy fiber terminals and in synaptic plasticity processes will be reviewed. Early in postnatal development, MFs release into their targeted neurons mainly GABA which is depolarizing and excitatory. Endogenous activation of GluK1 KARs localized on MF terminals by glutamate present in the extracellular space down regulates GABA release, leading sometimes to synapse silencing. The depressant effect of GluK1 on MF responses is mediated by a metabotropic process, sensitive to pertussis toxin and phospholipase C (PLC) along the transduction pathway downstream to G protein activation. Blocking PLC with the selective antagonist U73122, unmasks the potentiating effect of GluK1 on MF-evoked GABAergic currents, which probably depend on the ionotropic type of action of these receptors.

In addition, GluK1 KARs dynamically regulate the direction of spike-time dependent plasticity, a particular form of Hebbian type of learning which consists in bidirectional modifications in synaptic strength according to the temporal order of pre and postsynaptic spiking. At immature MF-CA3 synapses pairing MF stimulation with postsynaptic spiking and vice versa induces long term depression of MF-evoked GABAergic currents. In the case of positive pairing synaptic depression can be switched into spike-time dependent potentiation by blocking GluK1 KARs with UBP 302. The depressant action exerted by GluK1 KARs on

*Corresponding Author: Enrico Cherubini, Neurobiology Sector and IIT Unit, International School of Advanced Studies (SISSA), Ed. Q1 Area Science Park, S.S.14 Km 163.5; 34012 Basovizza (Trieste) Italy. Email: cher@sissa.it

Kainate Receptors: Novel Signaling Insights, edited by Antonio Rodríguez-Moreno and Talvinder S. Sihra.

MF responses would prevent the excessive activation of the CA3 associative network by the excitatory action of GABA early in postnatal development.

Introduction

Glutamate receptors, which belong to the ligand-gated ion channels family, are integral membrane proteins mediating fast excitatory transmission in the brain. On the basis of their molecular, pharmacological and biophysical properties, three distinct receptor classes have been identified. They have been named after their respective agonists: *N*-methyl *D*-aspartate (NMDA), α-amino-3-hydroxy-5-methyl-4-isoxazolepropionic acid (AMPA) and kainate receptors. Among these, kainate receptors (KARs) have received particular attention only in recent years, due to the relatively late development of selective pharmacological tools which have allowed their specific functions to be distinguished from those of AMPA receptors.[1] Moreover, the generation of transgenic mice lacking specific KAR subunits has provided additional clues for understanding their respective role in synaptic transmission.[2]

Molecular cloning has identified five different KAR subunits, each of molecular mass of ~100 kDa, referred in accord to the new nomenclature as GluK1-GluK5,[3] which can co-assemble in various combinations to form functional receptor channels permeable to cations. In addition, RNA editing and alternative splicing further increase the likelihood of pharmacological and physiological heterogeneity in KAR family. While GluK1-GluK3 (formerly GluR5-7), when expressed in heterologous systems can form homomeric kainate-gated ion channels,[4-6] GluK4 and GluK5 (former KA1 and KA2) do not form channels on their own but only when co-expressed with other subunits.[7,8] These subunits are often termed the "high affinity" subunits because of their low nanomolar affinity for the seaweed toxin kainic acid.

KAR subunits are expressed throughout the CNS where they are distributed not only on dendrites and postsynaptic membranes but also predominantly on nerve fibers and synaptic terminals.[9] At postsynaptic site these receptors carry at least in part current charges of synaptic responses, while at presynaptic sites they exert a powerful control on transmitter release.[9]

In this Chapter we will focus mainly on KARs expressed in the hippocampus. Here, activation of KARs localized on CA3 pyramidal cells by glutamate released from mossy fiber terminals (MFs) induces slow excitatory postsynaptic currents (EPSCs) whose amplitude has been found to be strictly dependent on the pattern of stimulation.[10,11] MF-evoked synaptic currents involve GluK2 receptor subunits as demonstrated by the loss of slow kainate-mediated EPSCs in hippocampal slices from GluK2 KO mice.[12,13] In contrast, GluK2 KO mice still exhibit slow kainate-mediated EPSCs on GABAergic interneurons, suggesting the involvement of other KAR subtypes. The loss of kainate-mediated synaptic responses in GluK1 or GluK1/GluK2 KO mice suggests the involvement of GluK1 subunits.[14] The slower deactivation kinetics of kainate-mediated synaptic currents with respect to the AMPA ones, would allow integrating excitatory inputs on the targeted neuron over a larger time window.

Presynaptic Kainate Receptors Control the Release of Glutamate and GABA

In addition to their postsynaptic role, KARs play a crucial role in regulating transmitter release at both excitatory at inhibitory synapses. In particular, at excitatory synapses, kainate-induced increase in glutamate release contributes to short- and long-term synaptic plasticity processes. While in the case of excitatory synapses, KARs are activated by glutamate released from glutamatergic nerve endings in the case of inhibitory synapses KARs are stimulated by glutamate present in the extracellular medium or spilled over from adjacent excitatory synapses. At excitatory synapses, KARs have been shown to facilitate glutamate release, particularly during frequency-dependent facilitation, a form of short-term plasticity characteristic of MF-CA3 synapses[15-17] (but see Kwon and Castillo).[18] Synaptic facilitation can be mimicked by low concentrations of kainate while high concentrations of the agonist produce synaptic depression.[19] The facilitatory effect of kainate has been attributed to kainate-induced depolarization of presynaptic boutons or axon terminals[16,20,21] and the release of calcium from local stores.[22] Similarly, KARs bi-directionally modify synaptic efficacy at inhibitory connections.[14,23-27] Here, the depressant effect of kainate on GABA release has been shown to be mediated by a different, noncanonical signaling modality, involving a metabotropic instead of a ionotropic type of action.[25] This may occur directly through a G-protein coupled KAR or indirectly through the participation of an intermediate or a "linker" protein. Whatever the case, this kind of response is usually sensitive to inhibitors of the putative metabotropic signaling pathway(s) thought to be involved. Thus, kainate-induced depression of IPSCs evoked in CA1 principal cells by stimulation of stratum oriens or stratum radiatum interneurons was found to be dependent on the activation of a pertussis toxin (PTX)-sensitive G protein and was suppressed by inhibitors of phospholipase C (PLC) and protein kinase C (PKC) but independent of presynaptic ion channel activity.[25] Furthermore, biochemical studies clearly demonstrated the involvement of $G_{i/o}$ proteins, PLC and PKC in kainate-induced depression of GABA release from hippocampal synaptosomes, which by definition are devoid of somato-dendritic compartments.[28,29] Altogether these data give direct support to the assumption that inhibition of GABA release by hetero-synaptically activated KARs localized on presynaptic GABAergic terminals occurs via a novel noncanonical metabotropic type of mechanism.

Similarly, at excitatory CA3-CA1, the depressant effect of kainate on glutamate release was demonstrated to be mediated by G-protein coupled KARs possibly through a direct interaction of βγ subunits of G proteins with presynaptic voltage-dependent calcium channels.[30] Unlike Schaffer collaterals, at mossy fibers (MF)-CA3 connections, the mechanisms involved in the depressant effect of high concentrations of kainate on glutamate release are still under debate. While early reports have attributed this effect to the ionotropic action of kainate which would induce a depolarization of presynaptic terminals with consequent inactivation of sodium, calcium channels and/or electrical shunting,[20,31,32] a more recent study has suggested a metabotropic type of mechanism.[33] In support of this hypothesis is the observation that the reduction of glutamate release by kainate was prevented by PTX, a $G_{i/o}$ inhibitor. In addition, the depressant effect of kainate on glutamate release from MF terminals was found to be dependent on the activity of adenylyl cyclase, cAMP/PKA signaling cascade.[33] Interestingly, KARs via a metabotropic type of action have been shown to contribute together with Type II mGluRs to LTD induced at MF-CA3 synapses by low frequency stimulation of afferent inputs.[34,35]

Kainate Receptors Modulate Cell Excitability via a Metabotropic Type of Action

Kainate, at nanomolar concentrations, is known to increase cell excitability by inhibiting at postsynaptic level the calcium-activated potassium current (I_{sAHP}),[36,37-39] which contributes to spike frequency adaptation. This effect has been attributed to the metabotropic action of KARs since it involves a PTX-sensitive G-protein and downstream signaling pathways including PKC and PKA.[37-39] The concentrations of kainate needed to block I_{sAHP} (IC_{50} values 15 and 5 nM in CA1 and CA3 pyramidal cells, respectively) are within the range of high affinity KARs binding sites and within the range of GluK4 and GluK5 KAR subunits, suggesting the involvement of these subunits in regulating cell excitability. As expected the inhibition of I_{sAHP} by KA was lost in GluK5 knock out mice.[39] However, KA-induced inhibition of I_{sAHP} was lost also in GluK2 knock out mice,[38,39] probably due to the indirect loss of GluK5 because in the absence of GluK2 subunits, GluK5 are likely to undergo degradation.[40] In other words, GluK5 subunits assembled in heteromeric complexes with GluK2 subunits, would be critical for triggering the activation of G proteins. In contrast with this view, a recent study has shown that in double GluK4 and GluK5 KO mice low concentrations of kainate are still able to inhibit I_{sAHP} suggesting that high affinity GluK4 and GluK5 receptors, thought to be essential for the metabotropic type of action of KARs are not obligatory.[13]

The way in which these two different modalities (ionotropic and metabotropic) interact is intriguing and not fully understood. In particular, the molecular mechanisms by which KARs operate in a metabotropic modality are still unclear given that the receptors do not exhibit the classical topology of metabotropic G-protein-coupled receptors. A recent study has outlined the possibility that the dual modes of signaling interact to dynamically auto-regulate the number of KAR expressed on the cell surface.[41] It has been demonstrated that the repetitive activation of KARs, expressed in a recombinant system, is able to trigger their internalization via a G-protein dependent PKC phosphorylation mechanism, thus limiting their over-activation.[41]

Kainate Receptors in the Immature Hippocampus

Ionotropic glutamate receptors play key roles in multiple developmental mechanisms, including regulation of neuronal migration, differentiation and synapses formation. KARs are developmentally regulated, given their temporal and spatial expression is transient in some brain regions, compatible with their crucial role in controlling the maturation of neuronal networks.[42] The developmental profile of KARs gene expression is well preserved during evolution: it can be detected also in *Drosophila* during a major period of neurogenesis.[43] In an early study on the developing mouse brain, Bettler et al[44] found that gene transcripts for the GluK1 subunits are expressed in the entire CNS already between E10 and E14, a period of intense cellular differentiation. GluK1 transcript levels which are particularly pronounced in areas where synaptogenesis is in progress later in development become more spatially restricted and downregulated. In the hippocampus, high affinity kainate binding sites can be detected already at E14. GluK1-5 genes expression undergoes a peak in the late embryonic/early postnatal period (in coincidence with periods of intense synaptogenesis) and starts declining at P14.[45,46] In particular, the perinatal peak of GluK1 gene expression observed in the barrel cortex, septum, thalamus and CA1 interneurons is compatible with a role of this subunit in developmental plasticity. The disruption of barrel cortex formation after chronic blockade of ionotropic glutamate receptors (during the first 24 hours after birth) further supports this hypothesis.[47] The decline in KAR subunits expression observed at late developmental stages could be related to cell elimination, due to network remodeling. The permeability of some edited forms of GluK1 and GluK2 KAR subunits to calcium may contribute to limit neuronal number via their excitotoxic action which leads to cell death. Altogether, these observations indicate that KARs are essential for neuronal development.

In the hippocampus, mossy fibers, the axons of dentate gyrus granule cells, which are endowed with presynaptic KARs,[48] develop postnatally producing three types of connections with their targets: (i) mossy terminals which synapse with thorny excrescences of CA3 principal cells; (ii) small en passant terminals; (iii) filopodial extensions which specifically contact GABAergic interneurons.[49] In a study aimed at elucidating the motility of axonal filopodia and dendritic spines during postnatal development, a process thought to be involved in synapse formation and rearrangement, it was demonstrated that synaptic activation of KARs enhances motility at mossy fiber axons in younger hippocampal slice culture but inhibits it in more mature slices,[50] suggesting a key role for KARs for establishing synaptic contacts and for stabilizing newly formed ones.

More extensive studies relate to the functional role that KARs exert on network activity in the immature hippocampus. It is worth noting that correlated neuronal activity constitutes a hallmark of developmental networks, well preserved during evolution that has been observed not only in the hippocampus but in almost every brain structure examined, including the retina,[51] the neocortex,[52-55] the hypothalamus,[56] the cerebellum[57] and the spinal cord.[58,59] In the hippocampus, the so-called giant depolarizing potentials or GDPs are generated by the synergistic action of synaptically released glutamate and GABA, both of which are depolarizing and excitatory.[60,61] GDPs which have been proposed to be the in vitro counterpart of "sharp waves" recorded in rat pups during immobility periods, sleep and feeding[62] can be considered a primordial form of synchrony between neurons, which precedes more organized forms of activity such as the theta and the gamma rhythms crucial for information processing.[63] GDPs disappear spontaneously towards the end of the second postnatal week in concomitance with the shift of GABA from the depolarizing to the hyperpolarizing direction. The strong developmental correlation between GDPs and KARs expression suggests that the latter are crucial for regulating early network activity which in turn contributes to sculpt neuronal circuits. In particular, signaling via presynaptic GluK1 containing KARs has been shown to be critical for regulating the number of functional glutamatergic synapses[64] and the balance between GABAergic and glutamatergic transmission which control GDPs.[27,65,66]

The hippocampal network comprises a large variety of distinct locally connected GABAergic interneurons which by pacing, timing and synchronizing principal cells give rise to coherent oscillations in both the adult[67] as well as in the immature brain.[68] Network synchronization is tightly controlled by the intrinsic properties of GABAergic interneurons[69] and by their electrical coupling via gap junctions.[70-72] Recent evidence indicates that tonic activation of GluK1 subunit-containing KARs by "ambient" glutamate can enhance, via a metabotropic action, the firing of neonatal interneurons by interfering with the calcium-dependent after hyperpolarization of medium duration (mAHP), generated by an apamin-sensitive K^+ current mediated by SK channels.[73] This effect is age-dependent since it disappears towards the end of the second postnatal week in concomitance with the uncoupling between KARs activation and mAHP. This mechanism together with the developmental shift of GABA from the depolarizing to the hyperpolarizing direction may contribute to the age-dependent disappearance of GDPs in the hippocampus.

Interestingly, at MF-CA3 connections presynaptic KARs are functional at early developmental stages,[27,66,67,74] while postsynaptic ones become operative only towards the end/beginning of the second postnatal week.[75] Thus, while in adulthood, stimulation of MF terminals evokes in CA3 principal cells excitatory postsynaptic currents (EPSCs) which carry at least in part current charges via KARs,[9,10,12,76] in neonates MFs terminals release mainly GABA (see below). A postsynaptic KAR-mediated component of the EPSC starts appearing during the second postnatal week when MFs acquire the classical glutamatergic phenotype in concomitance with the shift of GABA from the depolarizing to the hyperpolarizing direction.[75] The appearance of KAR-dependent EPSCs tightly coincides with the development of large amplitude AMPA-mediated EPSCs and with frequency-dependent facilitation. The lack of KARs-mediated postsynaptic components with summation properties which facilitate the temporal integration of synaptic inputs might change the way of synaptic signaling early in postnatal development. The lack of GluK1 and GluK2 KAR containing subunits would perturb the maturation of AMPA-mediated components of MF-EPSCs through pre and postsynaptic actions as assessed using mutant GluK1 and $GluK2^{-/-}$ mice.[75] Interestingly, GluK2 subunits of KARs interact with cell-adhesion molecules of the cadherin-catenin complex,[77] known to be involved in synapse formation together with the nectin-afadin system.[78] Therefore, the possibility that maturation of MF-CA3 synapses occurs via the mutual interaction of KARs with adhesion molecules represents an intriguing hypothesis.

GABAergic Phenotype of Immature MF Terminals

In adulthood, the primary excitatory neurotransmitter released from MF terminals is glutamate. However, in particular conditions, in addition to glutamate, MFs can release GABA. Thus, Gutiérrez and Heinemann[79] have convincingly demonstrated that in kindled rats monosynaptic bicuculline-sensitive GABAergic inhibitory postsynaptic potentials (IPSPs) occur in CA3 principal cells in response to stimulation of granule cells in the dentate gyrus. Interestingly, a high frequency input to pyramidal cells leads to IPSPs summation and consequent reduction in the probability of generating action potentials. It seems therefore likely that the release of GABA from MF terminals during seizures may act as a compensatory mechanism to counterbalance the enhanced excitability induced by the epileptic activity.[80] Interestingly, seizures have been shown to be associated with a transient up-regulation of presynaptic GABA ergic markers such as GAD_{67}, GAD_{65} and VGAT.[81-83] Moreover, hippocampal pyramidal neurons are able to express not only glutamate but also "mistargeted" $GABA_A$ receptors which, in particular conditions may become functional.[84] This suggests that MF can use GABA as a neurotransmitter since they posses all the machinery for synthesizing, storing, releasing and sensing it. Indeed, in juvenile animals, stimulation of granule cells in the dentate gyrus induces monosynaptic GABAergic and glutamatergic responses in CA3 principal cells[85,86] and immunogold experiments have demonstrated that AMPA and $GABA_A$ receptors are colocalized on the same synapse in close apposition to MF terminals.[87]

In contrast, immediately after birth, GABA is the main neurotransmitter released from MF terminals. At this early developmental stage, MFs constitutively express GAD_{67} and its product GABA[81,82] as well as the mRNA for the vesicular GABA transporter VGAT.[88] Post embedding immunogold double labeling has revealed the coexistence in MF terminals of VGAT and VGLUT (the vesicular transporter for glutamate) further suggesting that GABA can be coreleased with glutamate.[89]

Thus, minimal stimulation of granule cells in the dentate gyrus evokes in CA3 principal cells monosynaptic currents that completely fulfill the criteria for MF identification. These currents which are insensitive to AMPA receptor antagonists[74] but sensitive to the selective group III metabotropic glutamate receptor agonist L-(+)-2-Amino-4-phosphonobutyric acid (L-AP4), are readily blocked by picrotoxin, bicuculline or gabazine, suggesting that they are mediated by GABA acting on $GABA_A$ receptors (Fig. 1).

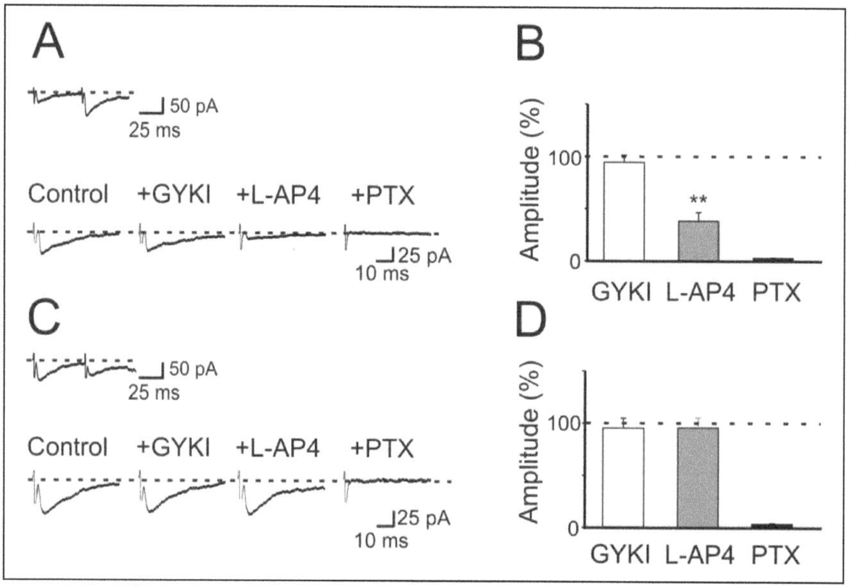

Figure 1. GABAergic origin of mossy fibers-evoked unitary postsynaptic currents in immature CA3 pyramidal cells. A) Unitary synaptic currents evoked in a CA3 pyramidal cell by minimal stimulation of granule cells in the dentate gyrus at P3 in control conditions, during application of GYKI 52466 (30 μM), GYKI plus L-AP4 (10 μM) and after addition of picrotoxin (PTX, 100 μM). Each trace is the average of 15-20 responses (including failures). Note that GYKI failed to modify synaptic currents. Synaptic currents were reduced in amplitude by L-AP4 and abolished by PTX. In the inset above the traces, MF-evoked GPSCs showing paired-pulse facilitation. B) Each column represents the mean peak amplitude current obtained in the experimental conditions shown in A and normalized to controls (dashed line; n = 6). In this and in the following figures, vertical bars refer to SEM. C) Example of GPSCs probably originated from a GABAergic interneuron impinging into a pyramidal cell exhibiting paired pulse depression. In this case, GPSCs were insensitive to L-AP4, but were blocked by PTX. D) As in B but for L-AP4 insensitive GABAergic interneurons (n = 7). $^{**}p < 0.01$.

As expected for $GABA_A$-mediated postsynaptic currents (GPSCs), synaptic responses can be potentiated by flurazepam, an allosteric modulator of $GABA_A$ receptors or by 1-(2-(((diphenylmethylene)amino)oxy)ethyl)-1,2,5,6-tetrahydro-3-pyridinecarboxylic acid (NO-711) a selective blocker of the GABA transporter GAT-1.[90] Additional evidence in favor of GABA as a neurotransmitter at MF-CA3 synapses is provided by the experiments in which, chemical stimulation of granule cell dendrites in stratum moleculare with glutamate (in the presence of the AMPA/kainate receptor antagonists to prevent the recruitment of GABAergic interneurons), depolarizes granule cells via activation of NMDA receptors and induces in CA3 principal cells barrages of GABAergic currents sensitive to L-AP4. Moving the pressure pipette few μm away towards the hilus to activate hilar interneurons causes barrage of $GABA_A$-mediated events insensitive to L-AP4, implying that they are mediated by the release of GABA from GABAergic interneurons.[90]

Perforated patch experiments, to preserve the anionic conditions of the recorded cells, have revealed that GABA released from MF terminals, exert a depolarizing action on targeted cells.[91] Accumulation of chloride inside the cell via the cation-chloride cotransporter NKCC1 is responsible for the depolarizing action of GABA since the positive driving force for chloride (+ 9 mV) observed in control conditions shifts towards negative value (-7 mV) when slices are exposed to the selective inhibitor of NKCC1, bumetanide,[92,93] thus confirming the general rule that GABA depolarizes immature neurons because of a reversed chloride gradient.[94]

In view of the GABAergic phenotype of immature MFs, what could the functional role of presynaptic KARs abundantly expressed on MF terminals? How could they be activated? How could they be involved in shaping neuronal networks? Some of these questions have been recently addressed.[74]

GluK1 KARs Down Regulate GABA Release by from Immature MF Terminals

In the immature hippocampus, activation of GluK1 KARs subunits by tonic glutamate present in the extracellular space has been shown to downregulate the release of GABA from MF terminals via a metabotropic type of action.[74] Thus, blocking GluK1 receptors with (S)-1-(2-Amino-2-carboxyethyl)-3-(2-carboxybenzyl)pyrimidine-2,4-dione (UBP 302), a specific GluK1 antagonist, in the presence of AMPA receptor blockers, enhances the amplitude of GPSCs (Fig. 2). As expected for a presynaptic type of action this effect is associated with an increase in successes rate, in CV^{-2} and a decrease in PPR (Fig. 2).

In addition, in "presynaptically" silent neurons, UBP 302 induces the appearance of synaptic responses to the first stimulus suggesting that endogenous activation of GluK1 contributes to synapses silencing (Fig. 3). Silent synapses represent a common feature of the developing brain.[95] They have been observed in a variety of different structures including the hippocampus where their number decreases significantly with age.[96]

The depressant action of GluK1 KARs on GPSCs is not indirectly mediated via other signaling molecules known to inhibit GABA release since KARs antagonists are still able to enhance the amplitude of GPSCs when applied in the presence of various receptor blockers including those for $GABA_B$, nicotinic, muscarinic, purinergic P2Y and mGlu.[74]

It is known that presynaptic receptors are usually activated by spillover of the neurotransmitter from axon terminals. One fundamental question to be addressed is how presynaptic GluK1 receptors can be activated if the main neurotransmitter released from MFs is GABA. One possibility is that these receptors are constitutively activated by ambient glutamate present in the extracellular medium which in neonates would be maintained at high levels by a less efficient glutamate transport mechanism and a poorly developed diffusional barrier.[97] Indeed, by enhancing the clearance of glutamate from the extracellular space, using an enzymatic glutamate scavenger system prevents the activation of presynaptic kainate receptors by glutamate and similarly to UBP 302 induces an increase in amplitude of GPSCs.[74]

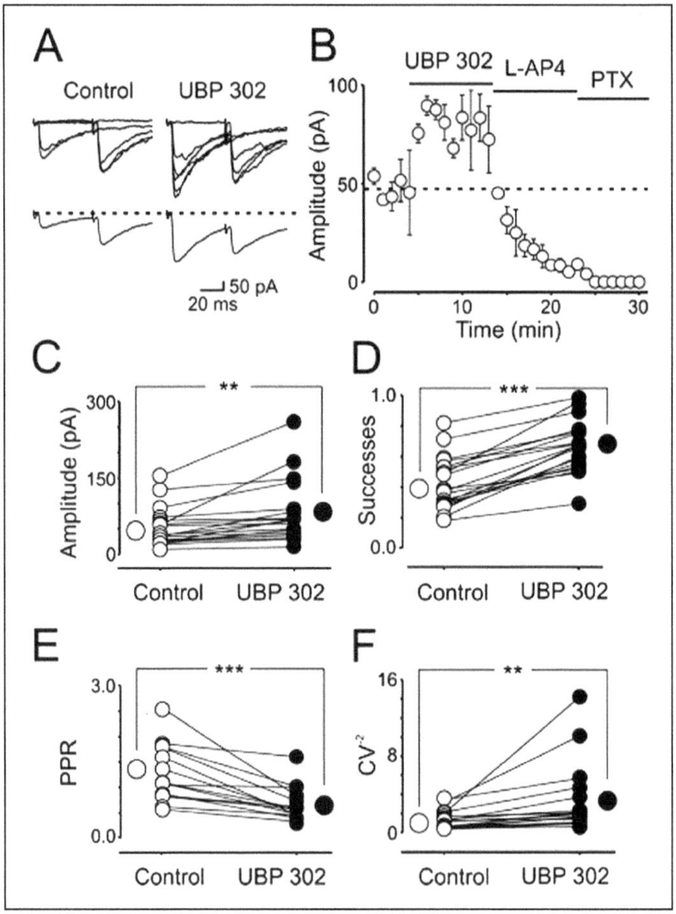

Figure 2. Endogenous activation of presynaptic kainate receptors down regulates MF-GPSCs. A) Superimposed individual traces of MF-GPSCs evoked in the presence of GYKI 52466 (30 μM, Control) and GYKI 52466 plus UBP 302 (10 μM). Below: averaged traces (successes plus failures). Note that UBP 302 enhanced the amplitude of the first response and reduced the number of synaptic failures. B) Summary plot showing the mean amplitude of GPSCs in the presence of GYKI and GYKI 52466 plus UBP 302 (n = 19). The horizontal dashed line refers to the mean amplitude value measured before UBP. C-F. Amplitude (C), Successes (D), PPR (E) and inversed square of CV (F) of MF-GPSCs measured in individual cells before and after application of UBP. Larger symbols represent averaged values. $^{**}p < 0.01$; $^{***}p < 0.001$. (Modified with permission from ref. 74.)

Since the depressant action of kainate on transmitter release seems to occur via G protein-coupled receptors,[19] to elucidate whether GluK1-induced depression of GPSCs operates through this particular modality, hippocampal slices were treated overnight with pertussis toxin (PTx) which blocks $G_{i/o}$ protein coupled receptors. In these conditions, UBP 302 fails to modify the amplitude of GPSCs. Similarly, the amplitude of GPSCs is unaltered when the glutamate scavenger is applied to slices incubated with PTx. The GluK1-induced-depression of MF-GPSCs involves G-coupled receptors localized on the presynaptic site, since blocking G

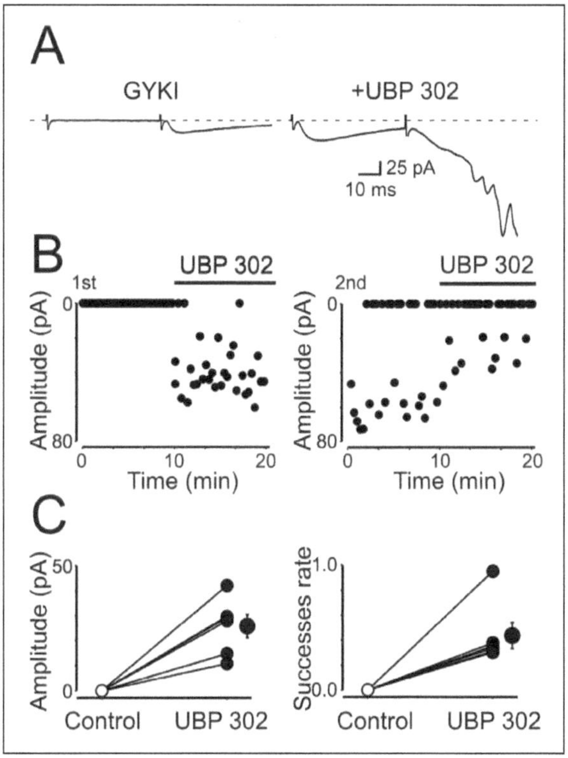

Figure 3. Blocking GluK1 receptors with UBP 302 enhances the probability of GABA release and converts "presynaptically" silent synapses into active ones. A) averaged traces of a "presynaptically" silent neuron recorded in the presence of GYKI and GYKI plus UPB 302. Note the appearance in the presence of UBP 302 of a synaptic current in response to the first stimulus. B) Time course of the peak amplitude of the first (left) and second (right) response shown in A, in control and during bath application of UBP (bars). C) Summary data (amplitude and successes rate) for 6 "presynaptically" silent cells.

proteins localized on the postsynaptic membrane using GDPβS into the patch pipette does not alter the action of UBP 302 on MF-mediated GABAergic currents, further indicating that the depression of MF-GPSCs is mediated by $G_{i/o}$ protein-coupled kainate receptors localized on MF terminals. Since the signaling pathway stimulated by G protein likely involves the release of calcium from intracellular stores by the activation of phospholipase C (PLC) and PKC stimulation, with consequent inhibition of voltage-dependent calcium channels,[25,98,99] disrupting the intracellular cascade downstream of G-protein activation with 1-[6-[[(17b)-3-Methoxyestra-1,3,5(10)-trien-17-yl]amino]hexyl]-1H-pyrrole-2,5-dione (U73122), a selective PLC blocker, should prevent the depressant action of GluK1 antagonist on GPSCs.

Interestingly, blocking PLC with U73122, downstream to G protein activation, unmasks the potentiating effect of GluK1 on MF GPSCs, which probably depends on the ionotropic type of action of this receptor. In keeping with this, the GluK1 antagonist UBP 302 significantly reduces the probability of evoking antidromic spikes in single granule cells by stimulation of MFs in stratum lucidum (Fig. 4).

A similar effect is produced by philanthotoxin which blocks calcium permeable AMPA/kainate receptors,[100] indicating that GluK1 increases MF excitability through the activation of

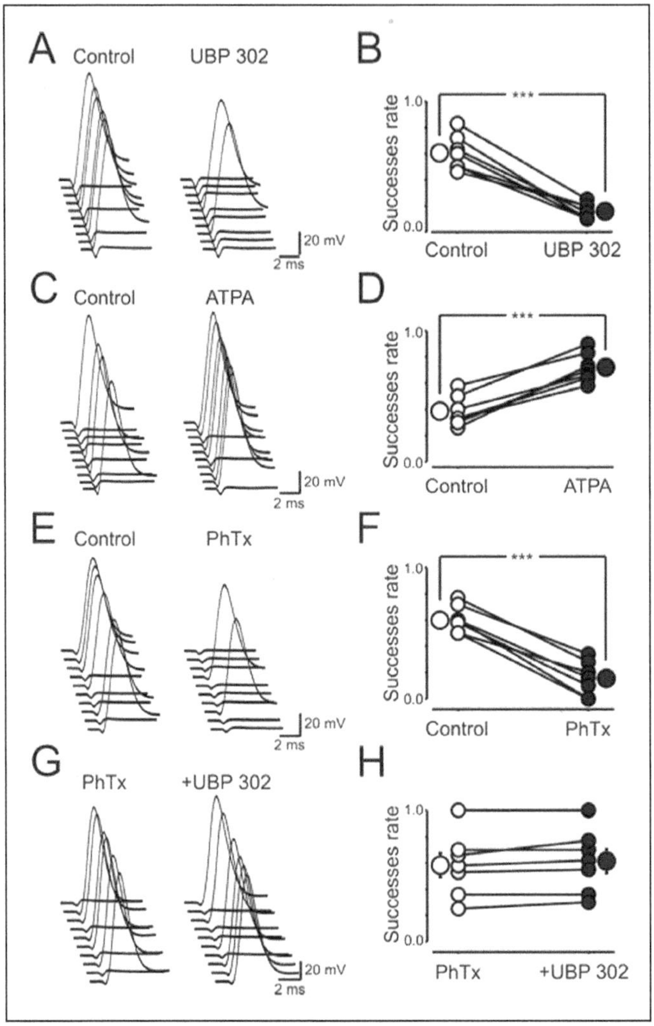

Figure 4. GluK1 receptors sensitive to philantotoxin control MF excitability. A) Consecutive traces showing antidromic spikes recorded in granule cells upon stimulation of MF in stratum lucidum before and during application of UBP 302 (note that the stimulus strength was set to obtain > than 50 % of successes. B) Summary plot of UBP 302 effects on successes rate (n = 7). C-D) In cells with < than 50% of successes, the selective GluK1 agonist ATPA enhanced MF excitability and the successes rate (n = 7). E-F) Philantotoxin (PhTx, 3 μM) mimicked the effects of UBP 302 (n = 7). G-H) In the presence of philantotoxin, UBP was not effective (n = 7). ****p* < 0.001. (Modified with permission from ref. 74.)

calcium-permeable cationic channels and depolarization of MF terminals (Fig. 4). Consistent with a ionotropic type of action, application of UBP 302 in the presence of philanthotoxin fails to modify MFs excitability (Fig. 4). In juvenile animals, activation of presynaptic kainate receptors has been found to directly depolarize via cation channels glutamatergic MF terminals[31] or GABAergic terminals,[101] thus lowering the threshold for antidromic action potential generation.

How can an increased MF excitability be reconciled with a depression of GABA release? According to Kamiya and Ozawa,[31] a down regulation of transmitter release may occur via inactivation of Na^+/Ca^{2+} channels or electrical shunting. However, this is unlikely in view of the recent finding that KA-induced facilitation of action potential evoked calcium entry in MF boutons involves a calcium store-dependent mechanism.[22] In addition, it should be stressed that unlike adults, immature MF terminate in very small spherical expansions[102] and do not exhibit use-dependent synaptic facilitation until the second week of postnatal life.[75] It is still unclear whether the dual signaling pathways (ionotropic and metabotropic), which depend on the common ionotropic GluK1 subunit, are independent or functionally coupled. In a previous study from dorsal root ganglion cells, it has been demonstrated that GluK1 KARs induces a G protein-dependent rise in $[Ca^{2+}]_i$, favoring its release from the internal stores.[98] It is tempting to speculate that calcium entering through calcium-permeable KARs may directly or indirectly interfere with G protein-mediated signaling leading to a dominant depressant effect on MF-GPSCs. The interplay between these two different pathways has been recently shown to account for the PKC-dependent autoregulation of membrane KARs.[41]

Modulation of Spike-Time Dependent Plasticity by GluK1 KARs

Activity-dependent changes in synaptic strength such as long-term-potentiation (LTP) or long-term depression (LTD) are critical for information storage in the brain and for the development of neuronal circuits. One interesting question is whether, early in postnatal development, MF-GPSCs can undergo activity-dependent modifications in synaptic efficacy. As already mentioned, immature neurons are characterized by an elevated number of silent synapses which can be converted into active ones by activity-dependent processes and this represents the most common mechanism for LTP induction, not only during development, but also in the mature brain.[95]

A form of synaptic plasticity extensively studied at glutamatergic synapses is spike-time dependent plasticity (STDP). This is a particular form of Hebbian type of learning which consists in bi-directional modifications of synaptic strength according to the temporal order of pre and postsynaptic spiking.[103] Thus, positively correlated pre and postsynaptic spiking (pre before post) within a critical time window leads to LTP whereas a negative correlation (post before pre) to LTD.

We used the STDP protocol to verify whether activity can modify the strength of GABAergic MF-CA3 connections.[91] Unlike conventional STDP, pairing (in current clamp mode) ten postsynaptic spikes (at 0.1 Hz) with unitary MF-GPSPs, persistently downregulates synaptic efficacy in a way that is independent of the temporal order of pre and postsynaptic stimulation (Fig. 5). The decrease in amplitude of GPSCs, which persists without decrement for periods of time variable from 40 to 60 min, reaches its maximum when in the case of positive pairing antidromic spikes follow MF stimulation with a delay of 15 ms (coincident with the peak of the synaptic potentials) or when, in the case of negative pairing, it precedes MF activation by 50 ms.

These effects, which require for their induction a rise of calcium in the postsynaptic cell via voltage-dependent calcium channels, are associated with a significant decrease in successes rate, in the inverse squared value of the coefficient of variation of responses amplitude and a significant increase in paired pulse ratio, suggesting a presynaptic site of expression.

Interestingly, GluK1 KARs control the direction of STDP at immature MF-CA3 synapses. Thus, when positive pairing (15 ms delay) is delivered in the presence of the selective GluK1 antagonist UBP 302 a shift from STD-LTD to STD-LTP occurs[91] (Fig. 5). In addition, this activity-dependent form of synaptic plasticity involves the activation of a G-protein and PLC since it can be prevented by the selective PLC blocker U73122.[74] However, in the presence of KAR antagonists, negative pairing (postsynaptic spiking preceding MF stimulation with a delay of 50 ms) still induces LTD similar in all respects to that obtained in the presence of a high concentration of DNQX which blocks both AMPA and kainate receptors,[91] suggesting that KAR are not involved in this form of synaptic plasticity (Fig. 5).

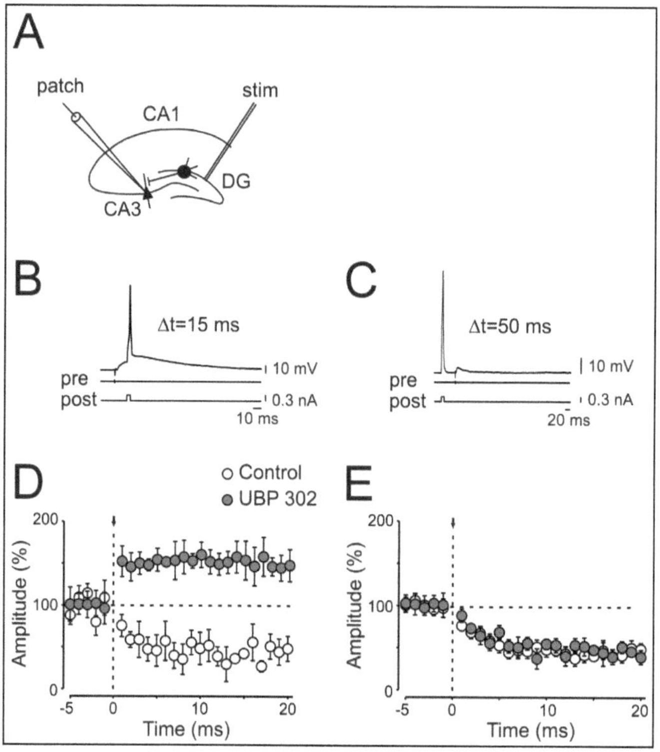

Figure 5. Presynaptic kainate receptors control the direction of spike-time dependent plasticity (STDP). A) Schematic representation of the experimental design. B) The stimulation of granule cells in the dentate gyrus (pre) preceded the postsynaptic spike (post) by 15 ms (Δt). C) the stimulation of granule cells in the dentate gyrus followed the postsynaptic spike by 50 ms. D) Summary plot of the mean peak amplitude of GPSCs recorded before and after positive pairing (arrow at time 0) in the absence (control, white symbols, n = 11) or in the presence of UBP 302 (grey symbols, n = 9). E) As in D but for negative pairing (post before pre; n = 7). Note the shift from spike-time dependent depression into spike-time dependent potentiation in the presence of UBP only in the case of positive pairing.

In summary, it is clear that tonic activation of presynaptic KAR by endogenous glutamate accounts for the persistent depression of MF-GPSCs observed after pairing presynaptic MF stimulation with postsynaptic spiking as demonstrated by the possibility to switch spike-time dependent depression into potentiation with UBP 302. Although the precise mechanisms underlying this phenomenon are still unclear, we cannot exclude the possibility that, with respect to LTD induced by negative pairing, KA-induced synaptic depression relies on a distinct calcium signal which in turn may activate a different molecular pathway, as suggested by the calcium hypothesis.[104]

Conclusion

Although much remains to be learned about the means by which KARs control GABA release at immature MF-CA3 connections and particularly the dual signaling pathways involved (ionotropic and metabotropic), it is conceivable that the depression of GABAergic transmission following tonic activation of GluK1 KARs by ambient glutamate represents a homeostatic mechanism which would limit the excessive activation of the auto-associative CA3 network by the excitatory action of GABA, thus preventing the occurrence of seizures. These properties are likely to be critical for information processing and for the proper development of the adult hippocampal circuitry.

Acknowledgements

This work was supported by a grant from Ministero Istruzione Universita' e Ricerca (MIUR) to EC.

References

1. Nisticò R, Dargan S, Fitzjohn SM et al. GLUK1 receptor antagonists and hippocampal mossy fiber function. Int Rev Neurobiol 2009; 85:13-27.
2. Huettner JE. Kainate receptors and synaptic transmission. Prog Neurobiol 2003; 70:387-407.
3. Collingridge GL, Olsen RW, Peters J et al. A nomenclature for ligand-gated ion channels. Neuropharm 2009; 56:2-5.
4. Egebjerg J, Bettler B, Hermans-Borgmeyer I et al. Cloning of a cDNA for a glutamate receptor subunit activated by kainate but not by AMPA. Nature 1991; 351:745-748.
5. Schiffer HH, Swanson GT, Heinemann SF. Rat GluR7 and a carboxy-terminal splice variant, GluR7b, are functional kainate receptor subunits with a low sensitivity to glutamate. Neuron 1997; 19:1141-1146.
6. Sommer B, Burnashev N, Verdoorn TA et al. A glutamate receptor channel with high affinity for domoate and kainate. EMBO J 1992; 11:1651-1656.
7. Herb A, Burnashev N, Werner P et al. The KA-2 subunit of excitatory amino acid receptors shows widespread expression in brain and forms ion channels with distantly related subunits. Neuron 1992; 8:775-785.
8. Werner P, Voigt M, Keinänen K et al. Cloning of a putative high-affinity kainate receptor expressed predominantly in hippocampal CA3 cells. Nature 1991; 351:742-744.
9. Lerma J. Roles and rules of kainate receptors in synaptic transmission. Nat Rev Neurosci 2003; 4:481-495.
10. Castillo PE, Malenka RC, Nicoll RA. Kainate receptors mediate a slow postsynaptic current in hippocampal CA3 neurons. Nature 1997; 388:182-186.
11. Vignes M, Collingridge GL. The synaptic activation of kainate receptors. Nature 1997; 388:179-182.
12. Mulle C, Sailer A, Pérez-Otaño I et al. Altered synaptic physiology and reduced susceptibility to kainate-induced seizures in GluR6-deficient mice. Nature 1998; 392:601-615.
13. Fernandes HB, Catches JS, Petralia RS et al. High-affinity kainate receptor subunits are necessary for ionotropic but not metabotropic signaling. Neuron 2009; 63:818-829.
14. Mulle C, Sailer A, Swanson GT et al. Subunit composition of kainate receptors in hippocampal interneurons. Neuron 2000; 28:475-484.
15. Contractor A, Swanson G, Heinemann SF. Kainate receptors are involved in short- and long-term plasticity at mossy fiber synapses in the hippocampus. Neuron 2001; 29:209-216.
16. Lauri SE, Bortolotto ZA, Bleakman D et al. A critical role of a facilitatory presynaptic kainate receptor in mossy fiber LTP. Neuron 2001; 32:697–709.
17. Pinheiro PS, Perrais D, Coussen F et al. GluR7 is an essential subunit of presynaptic kainate autoreceptors at hippocampal mossy fiber synapses. Proc Natl Acad Sci USA 2007; 104:12181-12186.
18. Kwon HB, Castillo PE. Role of glutamate autoreceptors at hippocampal mossy fiber synapses. Neuron 2008; 60:1082-1094.
19. Lerma J. Kainate receptor physiology. Curr Opin Pharmacol 2006; 6:89-97.
20. Kamiya H, Ozawa S, Manabe T. Kainate receptor-dependent short-term plasticity of presynaptic Ca^{2+} influx at the hippocampal mossy fiber synapses. J Neurosci 2002; 22:9237-9243.
21. Nicoll RA, Schmitz D. Synaptic plasticity at hippocampal mossy fibre synapses. Nat Rev Neurosci 2005; 6:863-876.
22. Scott R, Lalic T, Kullmann DM et al. Target-cell specificity of kainate autoreceptor and Ca^{2+}-store-dependent short-term plasticity at hippocampal mossy fiber synapses. J Neurosci 2008; 28:13139-13149.
23. Clarke VR, Ballyk BA, Hoo KH et al. A hippocampal GluR5 kainate receptor regulating inhibitory synaptic transmission. Nature 1997; 389:599-603.
24. Rodríguez-Moreno A, Herreras O, Lerma J. Kainate receptors presynaptically downregulate GABAergic inhibition in the rat hippocampus. Neuron 1997; 19:893-901.
25. Rodríguez-Moreno A, Lerma J. Kainate receptor modulation of GABA release involves a metabotropic function. Neuron 1998; 20:1211-1218.
26. Cossart R, Tyzio R, Dinocourt C et al. Presynaptic kainate receptors that enhance the release of GABA on CA1 hippocampal interneurons. Neuron 2001; 29:497-508.
27. Maingret F, Lauri SE, Taira T et al. Profound regulation of neonatal CA1 rat hippocampal GABAergic transmission by functionally distinct kainate receptor populations. J Physiol 2005; 567(Pt 1):131-142.
28. Cunha RA, Constantino MD, Ribeiro JA. Inhibition of [3H] gamma-aminobutyric acid release by kainate receptor activation in rat hippocampal synaptosomes. Eur J Pharmacol 1997; 323:167-172.

29. Cunha RA, Malva JO, Ribeiro JA. Pertussis toxin prevents presynaptic inhibition by kainate receptors of rat hippocampal [(3)H]GABA release. FEBS Lett 2000; 469:159-162.
30. Frerking M, Schmitz D, Zhou Q et al. Kainate receptors depress excitatory synaptic transmission at CA3→CA1 synapses in the hippocampus via a direct presynaptic action. J Neurosci 2001; 21:2958-2966.
31. Kamiya H, Ozawa S. Kainate receptor-mediated presynaptic inhibition at the mouse hippocampal mossy fibre synapse. J Physiol 2000; 523(Pt 3):653-665.
32. Schmitz D, Mellor J, Frerking M et al. Presynaptic kainate receptors at hippocampal mossy fiber synapses. Proc Natl Acad Sci USA 2001; 98:11003-11008.
33. Negrete-Díaz JV, Sihra TS, Delgado-García JM et al. Kainate receptor-mediated inhibition of glutamate release involves protein kinase A in the mouse hippocampus. J Neurophysiol 2006; 96:1829-1837.
34. Bortolotto ZA, Nistico R, More JC. Kainate receptors and mossy fiber LTP. Neurotoxicology 2005; 26:769-777.
35. Negrete-Díaz JV, Sihra TS, Delgado-García JM. Kainate receptor-mediated presynaptic inhibition converges with presynaptic inhibition mediated by Group II mGluRs and long-term depression at the hippocampal mossy fiber-CA3 synapse. J Neural Transm 2007; 114:1425-1431.
36. Gho M, King AE, Ben-Ari Y et al. Kainate reduces two voltage-dependent potassium conductances in rat hippocampal neurons in vitro. Brain Res 1986; 385:411-414.
37. Melyan Z, Wheal HV, Lancaster B. Metabotropic-mediated kainate receptor regulation of IsAHP and excitability in pyramidal cells. Neuron 2002; 34:107-114.
38. Fisahn A, Heinemann SF, McBain CJ. The kainate receptor subunit GluR6 mediates metabotropic regulation of the slow and medium AHP currents in mouse hippocampal neurones. J Physiol 2005; 562(Pt 1):199-203.
39. Ruiz A, Sachidhanandam S, Utvik JK et al. Distinct subunits in heteromeric kainate receptors mediate ionotropic and metabotropic function at hippocampal mossy fiber synapses. J Neurosci 2005; 25:11710-11718.
40. Vincent P, Mulle C. Kainate receptors in epilepsy and excitotoxicity. Neuroscience 2009; 158:309-323.
41. Rivera R, Rozas JL, Lerma J. PKC-dependent autoregulation of membrane kainate receptors. EMBO J 2007; 26:4359-4367
42. Miller LP, Johnson AE, Gelhard RE et al. The ontogeny of excitatory amino acid receptors in the rat forebrain—II. Kainic acid receptors. Neuroscience 1990; 35:45-51.
43. Ultsch A, Schuster CM, Laube B et al. Glutamate receptors of Drosophila melanogaster: cloning of a kainate-selective subunit expressed in the central nervous system. Proc Natl Acad Sci USA 1992; 89:10484-10488.
44. Bettler B, Boulter J, Hermans-Borgmeyer I et al. Cloning of a novel glutamate receptor subunit, GluR5: expression in the nervous system during development. Neuron 1990;5:583-595.
45. Bahn S, Volk B, Wisden W. Kainate receptor gene expression in the developing rat brain. J Neurosci 1994; 14:5525-5547.
46. Ritter LM, Vazquez DM, Meador-Woodruff JH. Ontogeny of ionotropic glutamate receptor subunit expression in the rat hippocampus. Brain Res Dev Brain Res 2002; 139:227-236
47. Schlaggar BL, Fox K, O'Leary DD. Postsynaptic control of plasticity in developing somatosensory cortex. Nature 1993; 364:623-626.
48. Represa A, Tremblay E, Ben-Ari Y. Kainate binding sites in the hippocampal mossy fibers: localization and plasticity. Neuroscience 1987; 20:739-748.
49. Acsády L, Kamondi A, Sík A et al. GABAergic cells are the major postsynaptic targets of mossy fibers in the rat hippocampus. J Neurosci 1998; 18:3386-3403.
50. Tashiro A, Dunaevsky A, Blazeski R et al. Bidirectional regulation of hippocampal mossy fiber filopodial motility by kainate receptors: a two-step model of synaptogenesis. Neuron 2003; 38:773-784.
51. Feller MB, Butts DA, Aaron HL et al. Dynamic processes shape spatiotemporal properties of retinal waves. Neuron 1997; 19:293-306.
52. Yuste R, Katz LC. Control of postsynaptic Ca^{2+} influx in developing neocortex by excitatory and inhibitory neurotransmitters. Neuron 1991; 6:333-344.
53. Owens DF, Boyce LH, Davis MB et al. Excitatory GABA responses in embryonic and neonatal cortical slices demonstrated by gramicidin perforated-patch recordings and calcium imaging. J Neurosci 1996; 16:6414-6423.
54. Dammerman RS, Flint AC, Noctor S et al. An excitatory GABAergic plexus in developing neocortical layer 1. J Neurophysiol 2000; 84:428-434.
55. Maric D, Liu QY, Maric I et al. GABA expression dominates neuronal lineage progression in the embryonic rat neocortex and facilitates neurite outgrowth via GABA(A) autoreceptor/Cl- channels. J Neurosci 2001; 21:2343-2360.
56. Chen G, Trombley PQ, van den Pol AN. Excitatory actions of GABA in developing rat hypothalamic neurones. J Physiol 1996; 494(Pt 2):451-464.

57. Eilers J, Plant TD, Marandi N et al. GABA-mediated Ca^{2+} signalling in developing rat cerebellar Purkinje neurones. J Physiol 2001; 536(Pt 2):429-437.
58. Wang J, Reichling DB, Kyrozis A et al. Developmental loss of GABA- and glycine-induced depolarization and Ca^{2+} transients in embryonic rat dorsal horn neurons in culture. Eur J Neurosci 1994; 6:1275-1280.
59. O'Donovan MJ. The origin of spontaneous activity in developing networks of the vertebrate nervous system. Curr Opin Neurobiol 1999; 9:94-104.
60. Cherubini E, Gaiarsa JL, Ben-Ari Y. GABA: an excitatory transmitter in early postnatal life. Trends Neurosci 1991; 14:515-519.
61. Ben-Ari Y. Excitatory actions of gaba during development: the nature of the nurture. Nat Rev Neurosci 2002; 3:728-739.
62. Leinekugel X, Khazipov R, Cannon R et al. Correlated bursts of activity in the neonatal hippocampus in vivo. Science 2002; 296:2049-2052.
63. Buzsáki G, Draguhn A. Neuronal oscillations in cortical networks. Science 2004; 304:1926-1929.
64. Vesikansa A, Sallert M, Taira T et al. Activation of kainate receptors controls the number of functional glutamatergic synapses in the area CA1 of rat hippocampus. J Physiol 2007; 583(Pt 1):145-157.
65. Lauri SE, Segerstråle M, Vesikansa A et al. Endogenous activation of kainate receptors regulates glutamate release and network activity in the developing hippocampus. J Neurosci 2005; 25:4473-4484.
66. Lauri SE, Vesikansa A, Segerstråle M et al. Functional maturation of CA1 synapses involves activity-dependent loss of tonic kainate receptor mediated inhibition of glutamate release. Neuron 2006; 50:415-429.
67. McBain CJ, Fisahn A. Interneurons unbound. Nat Rev Neurosci 2001; 2:11-23.
68. Ben-Ari Y, Khalilov I, Represa A et al. Interneurons set the tune of developing networks. Trends Neurosci 2004; 27:422-427.
69. Maccaferri G, McBain CJ. Long-term potentiation in distinct subtypes of hippocampal nonpyramidal neurons. J Neurosci 1996; 16:5334-5343.
70. Hestrin S, Galarreta M. Electrical synapses define networks of neocortical GABAergic neurons. Trends Neurosci 2005; 28:304-309.
71. Zsiros V, Maccaferri G. Electrical coupling between interneurons with different excitable properties in the stratum lacunosum-moleculare of the juvenile CA1 rat hippocampus. J Neurosci 2005; 25:8686-8695.
72. Minneci F, Janahmadi M, Migliore M et al. Signaling properties of stratum oriens interneurons in the hippocampus of transgenic mice expressing EGFP in a subset of somatostatin-containing cells. Hippocampus 2007; 17:538-553.
73. Segerstråle M, Juuri J, Lanore F et al. High firing rate of neonatal hippocampal interneurons is caused by attenuation of afterhyperpolarizing potassium currents by tonically active kainate receptors. J Neurosci 2010; 30:6507-6514.
74. Caiati MD, Sivakumaran S, Cherubini E. In the developing rat hippocampus, endogenous activation of presynaptic kainate receptors reduces GABA release from mossy fiber terminals. J Neurosci 2010; 30:1750-1759.
75. Marchal C, Mulle C. Postnatal maturation of mossy fibre excitatory transmission in mouse CA3 pyramidal cells: a potential role for kainate receptors. J Physiol 2004; 561(Pt 1):27-37.
76. Cossart R, Epsztein J, Tyzio R et al. Quantal release of glutamate generates pure kainate and mixed AMPA/kainate EPSCs in hippocampal neurons. Neuron 2002; 35:147-159.
77. Coussen F, Normand E, Marchal C et al. Recruitment of the kainate receptor subunit glutamate receptor 6 by cadherin/catenin complexes. J Neurosci 2002; 22:6426-6436.
78. Mizoguchi A, Nakanishi H, Kimura K et al. Nectin: an adhesion molecule involved in formation of synapses. J Cell Biol 2002; 156:555-565.
79. Gutiérrez R, Heinemann U. Kindling induces transient fast inhibition in the dentate gyrus-CA3 projection. Eur J Neurosci 2001; 13:1371-1379.
80. Jaffe DB, Gutiérrez R. Mossy fiber synaptic transmission: communication from the dentate gyrus to area CA3. Prog Brain Res 2007; 163:109-132
81. Schwarzer C, Sperk G. Hippocampal granule cells express glutamic acid decarboxylase-67 after limbic seizures in the rat. Neuroscience 1995; 69:705-709.
82. Sloviter RS, Dichter MA, Rachinsky TL et al. Basal expression and induction of glutamate decarboxylase and GABA in excitatory granule cells of the rat and monkey hippocampal dentate gyrus. J Comp Neurol 1996; 373:593-618.
83. Lamas M, Gómez-Lira G, Gutiérrez R. Vesicular GABA transporter mRNA expression in the dentate gyrus and in mossy fiber synaptosomes. Brain Res Mol Brain Res 2001; 93:209-214.
84. Rao A, Cha EM, Craig AM. Mismatched appositions of presynaptic and postsynaptic components in isolated hippocampal neurons. J Neurosci 2000; 20:8344-8353.
85. Walker MC, Ruiz A, Kullmann DM. Monosynaptic GABAergic signaling from dentate to CA3 with a pharmacological and physiological profile typical of mossy fiber synapses. Neuron 2001; 29:703-715.

86. Gutierrez R, Romo-Parra H, Maqueda J et al. Plasticity of the GABAergic phenotype of the "glutamatergic" granule cells of the rat dentate gyrus. J Neurosci 2003; 23:5594-5598.
87. Bergersen L, Ruiz A, Bjaalie JG et al. GABA and GABAA receptors at hippocampal mossy fibre synapses. Eur J Neurosci 2003; 18:931-941.
88. Gómez-Lira G, Lamas M, Romo-Parra H et al. Programmed and induced phenotype of the hippocampal granule cells. J Neurosci 2005; 25:6939-6946.
89. Zander JF, Münster-Wandowski A, Brunk I et al. Synaptic and vesicular coexistence of VGLUT and VGAT in selected excitatory and inhibitory synapses. J Neurosci 2010; 30:7634-7645.
90. Safiulina VF, Fattorini G, Conti F et al. GABAergic signaling at mossy fiber synapses in neonatal rat hippocampus. J Neurosci 2006; 26:597-608.
91. Sivakumaran S, Mohajerani MH, Cherubini E. At immature mossy-fiber–CA3 synapses, correlated presynaptic and postsynaptic activity persistently enhances GABA release and network excitability via BDNF and cAMP-dependent PKA. J Neurosci 2009; 29:2637-2647.
92. Dzhala VI, Talos DM, Sdrulla DA et al. NKCC1 transporter facilitates seizures in the developing brain. Nat Med 2005; 11:1205-1213.
93. Sipilä ST, Schuchmann S, Voipio J et al. The cation-chloride cotransporter NKCC1 promotes sharp waves in the neonatal rat hippocampus. J Physiol 2006; 573(Pt 3):765-773.
94. Ben-Ari Y, Gaiarsa JL, Tyzio R et al. GABA: a pioneer transmitter that excites immature neurons and generates primitive oscillations. Physiol Rev 2007; 87:1215-1284.
95. Voronin LL, Cherubini E. 'Deaf, mute and whispering' silent synapses: their role in synaptic plasticity. J Physiol 2004; 557(Pt 1):3-12.
96. Durand GM, Kovalchuk Y, Konnerth A. Long-term potentiation and functional synapse induction in developing hippocampus. Nature 1996; 381:71-75.
97. Danbolt NC. Glutamate uptake. Prog Neurobiol 2001; 65:1-105.
98. Rozas JL, Paternain AV, Lerma J. Noncanonical signaling by ionotropic kainate receptors. Neuron 2003; 39:543-553.
99. Rodríguez-Moreno A, Sihra TS. Metabotropic actions of kainate receptors in the CNS. J Neurochem 2007; 10:2121-2135.
100. Fletcher EJ, Lodge D. New developments in the molecular pharmacology of alphaamino-3-hydroxy-5-methyl-4-isoxazole propionate and kainate receptors. Pharmacol Ther 1996; 70:65-89.
101. Semyanov A, Kullmann DM. Kainate receptor-dependent axonal depolarization and action potential initiation in interneurons. Nat Neurosci 2001; 4:718-723.
102. Amaral DG, Dent JA. Development of the mossy fibers of the dentate gyrus: I. A light and electron microscopic study of the mossy fibers and their expansions. J Comp Neurol 1981; 195:51-86.
103. Dan Y, Poo MM. Spike timing-dependent plasticity: from synapse to perception. Physiol Rev 2006; 86:1033-1048.
104. Caporale N, Dan Y. Spike timing dependent plasticity: a Hebbian learning rule. Annu Rev Neurosci 2008; 31:25-46.

Chapter 3

Localization and Functions of Kainate Receptors in the Basal Ganglia

Xiao-Tao Jin* and Yoland Smith

Abstract

Kainate receptors (KARs) are one of the three subtypes of ionotropic glutamate receptors in the CNS. These receptors are widely expressed pre- and postsynaptically throughout the brain. Thus, kainate receptor activation mediates a large variety of pre- and postsynaptic effects on either glutamatergic or GABAergic synaptic transmission. Although ionotropic functions for KAR have been described in multiple brain regions, there is considerable evidence from various CNS regions that KARs activation modulates GABA release through either G-protein dependent metabotropic pathway or secondary activation of G-protein coupled receptors. In the present chapter, we provide further evidence supporting that these two pathways are also involved in the modulation of GABA release in specific basal ganglia nuclei. Because of their more subtle effects on neurotransmisison regulation than other ionotropic glutamate receptors, KARs represent interesting targets for the future development of pharmacotherapy for basal ganglia diseases.

Introduction

Kainate receptors (KARs) are one of the three subtypes of ionotropic glutamate receptors in the CNS, made up of a combination of GluR5, GluR6, GluR7, KA1 and KA2 subunits.[1,2] These receptors are widely expressed pre- and postsynaptically throughout the brain.[3-9] Cloning technology and recent development of drugs that could discriminate between kainate and AMPA receptors have led to further characterization of the pharmacological and physiological properties of KARs during the past decade.[7,8] Kainate receptors have direct actions on intrinsic cell excitability in the hippocampus[10] and modulate both glutamatergic and GABAergic synaptic transmission in a number of brain regions.[7,8] In the hippocampus and amygdala, synaptically released glutamate from nearby excitatory synapses, can activate KARs in GABAergic terminals, thereby providing an endogeneous, physiologically relevant, mechanism mimicking the effect of exogenous KA application on inhibitory synaptic transmission.[11,12] Although the hippocampal KARs surely deserved most attention, data from our laboratory and others have provided further evidence for a widespread pre and postsynaptic localization of KARs and their functions in many other CNS structures, including the cerebral cortex, hypothalamus, cerebellun, spinal cord and basal ganglia.[5,7,9,13] In this chapter, we will provide an overview of our current understanding of the localization and function of KARs in the basal ganglia and discuss their potential relevance as novel targets for movement disorders therapy. Each basal ganglia nucleus will be discussed in turn, followed by concluding remarks related to the potential impact KARs may have in regulating basal ganglia function under normal and pathological conditions. Additional information about the basal ganglia KARs localization can also be found in a previous review.[14]

*Corresponding Author: Xiao-Tao Jin—Division of Neuroscience, Yerkes National Primate Research Center and Department of Neurology, Emory University, 954, Gatewood Road Northeast, Atlanta, Georgia 30322 USA. Email: jxiaota@emory.edu

Kainate Receptors: Novel Signaling Insights, edited by Antonio Rodríguez-Moreno and Talvinder S. Sihra.

Striatum

The mRNA for GluR6, GluR7 and KA2, but not GluR5 and KA1, are expressed in rat and mice striatum.[15,16] Double in situ hybridization studies dermonstrated that most striatal projection neurons labeled for either enkephalin or substance P mRNAs co-express GluR6 mRNA.[17] The expression of KARs subunits in striatal interneurons remains to be determined. Consistent with these results, strong cellular and neuropil GluR6/7 immunoreactivity was found in the monkey striatum.[3,5] At the ultrastructural level, both GluR6/7 and KA2 are expressed in subsets of glutamatergic terminals, some of which in the sensorimotor putamen, originate from the motor cortex or the centromedian thalamic nucleus[5] (Fig. 1). Presynaptic KARs are either extrasynaptic, away from the main release sites

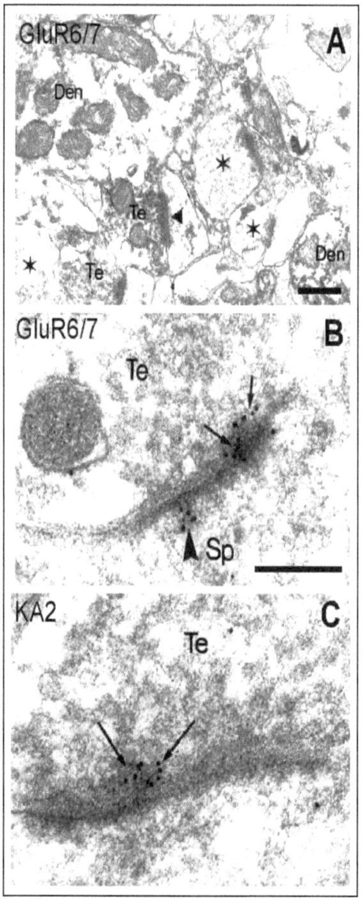

Figure 1. Pre and postsynaptic expression of GluR6/7 and KA2 immunoreactivity in the monkey striatum. A) GluR6/7-immunoreactive elements in the body of the caudate nucleus. Note the presence of labeled terminals (Te) forming asymmetric synapses (arrowheads) and immunoreactive dendrites (*Den*). The asterisks indicate unlabeled boutons. B) Presynaptic and postsynaptic GluR6/7 labeling at an asymmetric axospinous synapse as reviewed with the postembedding immunogold method. C) Dense KA2 labeling (arrows) in the presynaptic gird of an asymmetric axospinous synapse. Scale bars: A: 0.3 μm; B: 0.2 μm (valid for C). (See ref. 5 for more details). From: Kieval JZ et al. J Neurosci 2001; 21:8746-8757;[5] © 2001 with permission from the Society for Neuroscience.

of neurotransmitter or associated with the presynaptic active zones of glutamatergic synapses (Fig. 1).[5] At the postsynaptic level, the majority of GluR6/7 and KA2 labeling is found extrasynaptically along the plasma membrane of spines and dendrites contacted by glutamatergic inputs, though some labeling can occasionally be found in the postsynaptic density of glutamatergic synapses (Fig. 1).[5]

Agonist-induced activation of KARs in slices of rat striatum modulates GABAergic transmission.[17,18] Domoate (200 nM-500 nM), an AMPA/KAR agonist, increases the frequency of spontaneous GABAergic IPSCs (SIPSCs) with low amplitude, which are likely generated by intrinsic axon collaterals of GABAergic projection neurons,[20,21] but does not have any effect on the frequency of large amplitude IPSCs most likely mediated by GABAergic afferents from fast spiking striatal interneurons[19-21] or projections from the globus pallidus.[22-25] Domoate also decreases the IPSCs amplitude evoked by intrastriatal stimulation.[17] These effects are mediated by activation of GluR6-containing KARs because they are lost in GluR6-deficient mice.[17] Thus, these differential effects of KARs agonist on sIPSC of different amplitude suggest a variable degree of expression and function of presynaptic KARs on intrinsic GABAergic projections from striatal output neurons or interneurons. The lack effect of domoate on large amplitude sIPSCs suggests that low concentrations of domoate do not trigger spike discharge in GABAergic interneurons.

As described in hippocampus and spinal cord,[26,27] the KAR-induced depression of GABAergic transmission in the striatum involves activation of secondary G-protein coupled receptors.[17] Domoate-induced inhibition of evoked IPSCs (eIPSCs) is, indeed, significantly reduced by A_{2A} receptor antagonists (Fig. 2). These observations, combined with electron microscopic evidence for the localization of A_{2A} receptor immunoreactivity in intrastriatal GABAergic axon collaterals of striatopallidal neurons,[28,29] provide a substrate whereby KARs could regulate GABAergic transmission via indirect activation of presynaptic A_{2A} receptors. However, KARs activation does not affect the frequency of miniature IPSCs (mIPSCs), while increasing that of evoked and spontaneous IPSCs, in the rat nucleus accumbens, suggesting the involvement of postsynaptic KARs.[18] Together, these studies provide solid evidence that KARs are clearly involved in the regulation of inhibitory transmission in both the dorsal and ventral striatum, but significant work remains to be done to elucidate the exact mechanisms underlying these effects. The role of KARs on GABAergic interneurons must also be examined carefully.

Globus Pallidus

Globus pallidus neurons express strong mRNA for GluR6 and KA2 in rodents.[16] Consistent with these findings, neurons in the external (GPe) and internal (GPi) pallidal segments in monkeys display moderate to strong GluR6/7 and KA2 immunoreactivity.[30] At the electron microscopic level, GluR6/7 labeling is expressed both postsynaptically in dendrites of pallidal neurons and presynaptically in GABAergic striatal terminals and putative glutamatergic terminals.[9,30] In addition, significant GluR6/7 immunoreactivity is expressed in unmyelinated axons throughout both pallidal segments suggesting a presynaptic role for KARs in this brain region.[9,30] The pattern of KARs immunoreactivity in the monkey pallidum is very similar to that found in both adult and young rats GP (Fig. 3).

In line with these immunocytochemical data, KA application inhibits GABAergic synaptic transmission through presynaptic mechanisms in slices of rat GP (Fig. 4). On the other hand, KA (1 μM) does not have a signifficant effect on whole cell resistance of rat GP neurons[9] suggesting that the KAR-induced depression of eIPSCs is not due to postsynaptic changes of the passive membrane properties of GP cells.

As discussed in other chapters, since the pioneer publication of Rodriguez-Moreno and Lerma in 1998,[31] various complex mechanisms involving G-proteins have been demonstrated by which KARs modulate GABAergic synaptic transmission in the hippocampus and other brain regions (reviewed in refs. 7,8,32-34). We recently studied the effect of another G-protein inhibitor *N*-ethylmaleimide (NEM) on KA-induced inhibition of GABAergic transmission in slices of rat GP and found that this G-protein antagonist is capable of blocking presynaptic KAR-induced inhibition of glutamatergic transmission in this brain region.[9] We also demonstrated that the

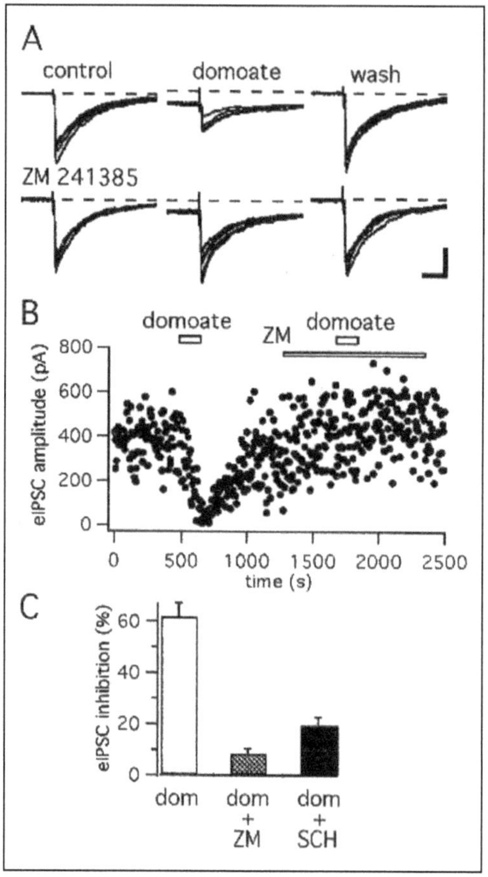

Figure 2. A_{2A} receptor antagonists block the action of domoate on evoked IPSCs. A) IPSCs were evoked by intrastriatal stimulation (at a rate of 0.2 Hz) in the presence of NBQX (1 μM). Top traces, Domoate (500 nM for 2 min) reversibly decreased evoked IPSC amplitude. Bottom traces, In the same cell, perfusion of the slice with the selective A_{2A} antagonist ZM 241385 (1 μM) prevents the action of domoate on evoked IPSC amplitude. The dotted line represents the level of the control inward current. B) For the same experiment, plot of the amplitude of evoked IPSCs as a function of the time. Domoate is applied at the time indicated by the open horizontal bars. As indicated by the bottom horizontal bar, ZM 241385 (1 μM) is perfused several minutes before the second application of domoate. C) Histogram of the average (± SEM) inhibition by domoate (500 nM for 2 min) of evoked IPSC amplitude in control condition (n = 18), in the presence of ZM 241385 (1 μM) (n = 7) and in the presence of SCH-58261 (1 μM) (n = 6). For both antagonists, the difference in inhibition versus domoate was significant with $P < 0.001$. Reprinted from Chergui K et al. J Neurosci 2000; 20(6):2175-2182;[17] © 2000 with permission from the Society for Neuroscience.

KAR-mediated presynaptic modulation of GABAergic synaptic transmission in GP is abolished by NEM (Fig. 5A,B), likely though a presynaptic site of action because the KAR-mediated decrease in mIPSCs frequency is also significantly reduced by this drug (Fig. 5C-E). Although the exact mechanisms by which G-proteins contribute to the KAR-mediated presynaptic effects in GP remain to be determined, the fact that these effects are not abolished after blockade of various G

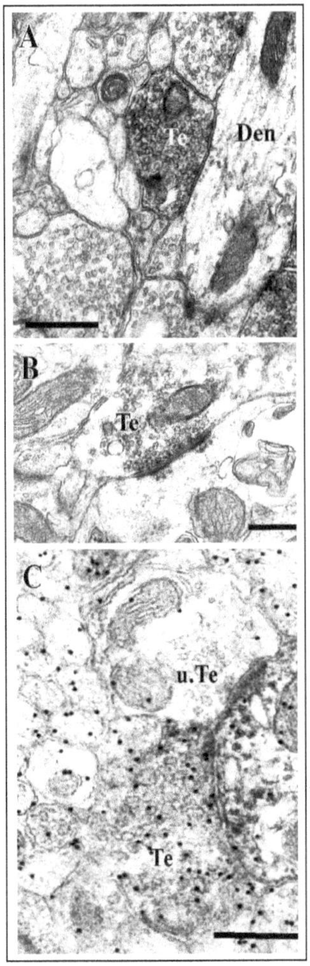

Figure 3. Pre and postsynaptic expression of GluR6/7 immunoreactivity in rat and monkey GP. GluR6/7-labeled terminals (*Te*) forming symmetric (A,C) or asymmetric (B) axon-dendritic synapse in rat and monkey pallidum. C) shows a GluR6/7-Postive axon terminals (Te) enriched in GABA immunoreactivity forming a symmetric axon-dendritic synapse on a labeled dendrite. Note the low density of gold particles associated with a putative glutamatergic terminal (u.Te) that forms an asymmetric synapse (arrow) on the same dendrite. Scale bars: A: 0.5 μm; B: 0.3 μm; C: 0.3 μm. (See references 9 and 30 for more details.) Reprinted from: Kane-Jackson R, Smith Y. Neuroscience 2003; 120:285-289;[30] with permission from Elsevier; and from Jin X-T et al. Eur J Neurosci 2006; 23:374-386;[9] © 2006 with permission from Wiley-Blackwell.

protein-coupled receptors, including $GABA_B$ and A_{2A} which are involved in mediated presynaptic effects of KARs in hippocampus and striatum,[17,26,27,35] is strongly indicative of a specific mechanism different from that seen at other synapses.[13]

Several studies have indicated that PKC activation downstream of G-protein activity is essential for KAR-mediated pre and postsynaptic effects (reviewed by refs. 33,34). For instance, KA-induced presynaptic inhibition of GABAergic or glutamatergic transmission and KA-mediated

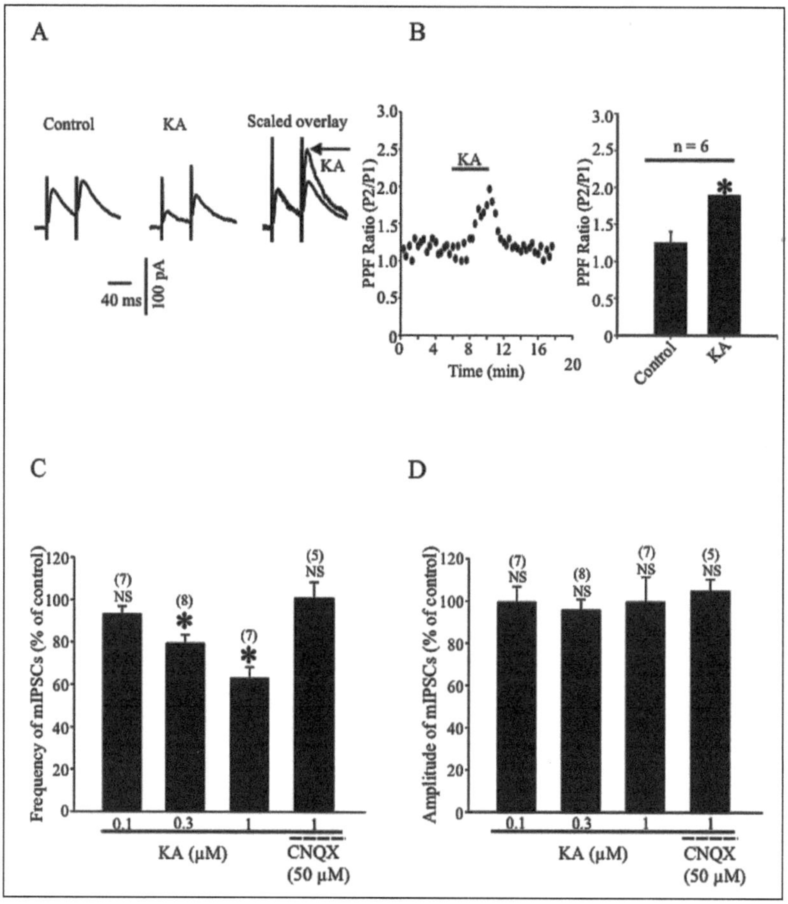

Figure 4. KAR activation increases paired pulse facilitation ratio (PPFR) and reduces the frequency, but not the amplitude, of mIPSCs at GABAergic synapses in the GP. A) Paired IPSCs were recorded before (left trace) and during (middle) 1 μM KA application. The right trace shows the KA-induced effect after scaling to the peak of the first IPSC. B) The same neuron presented in (A) shows the time course of increased paired-pulse facilitation ratio (PPFR) of IPSCs in response to 1 μM KA application (left graph) and the effect of KA on PPFR expressed as a ratio of P2/P1 (mean ± S.E.M.) (right graph). C) A summary bar graph shows that KA (0.1-0.3 μM) significantly reduces the frequency of mIPSCs, which is blocked in the presence of 50 μM CNQX. D) A summary bar graph shows that neither KA nor KA together with CNQX affects the amplitude of mIPSCs. Asterisks indicate a significant difference from control (* < 0.01), NS indicates nonsignificant differences; and n indicates the number of cells tested under each condition. (See ref. 13 for more details.) From X-T Jin, Smith Y et al. Neuroscience 2007; 149:338-349[13] © 2007 with permission from Elsevier.

postsynaptic inhibition of slow afterhyperpolarization currents (I_{sAHP}) in hippocampal neurons are blocked by PKC inhibitor (Calphostin C), but not by PKA inhibitor (H-89).[9,31,36-40] In the rat GP, presynaptic KAR-mediated effects on evoked and mIPSCs are also blocked by Calphostin, but not by H-89 (Fig. 6), providing further evidence that KAR-induced depression of GABAergic synaptic transmission in the rat GP requires G-protein and PKC activation, but does not rely on

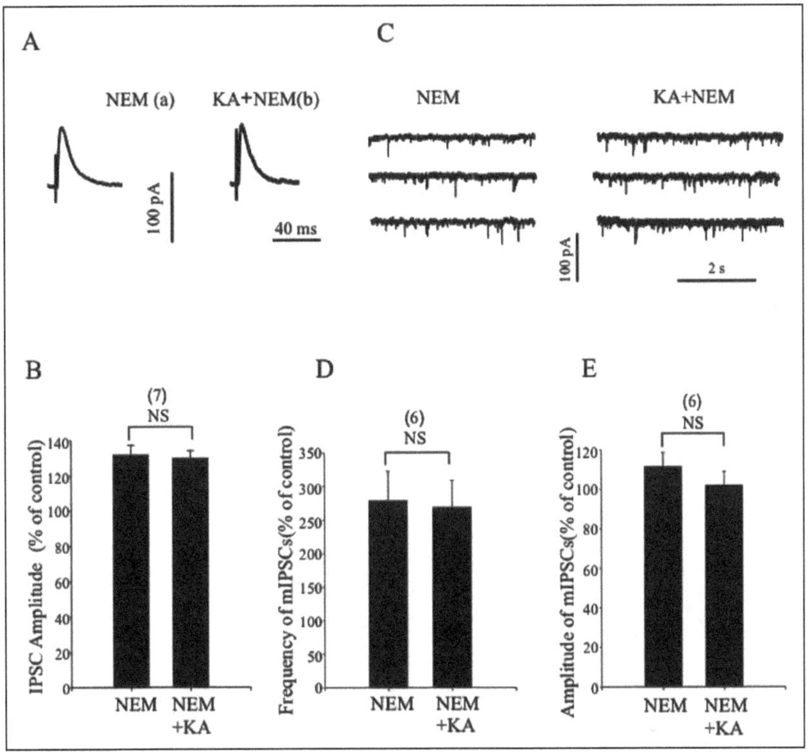

Figure 5. Application of G-protein inhibitor (NEM) blocks the KAR activation-induced inhibition of GABAergic synaptic transmission in rat GP. A) Evoked IPSCs recorded in presence of NEM (left trace) and NEM together with KA (right trace). B) A summary bar graph shows that the inhibitory effect of KA on IPSC amplitude is blocked in the presence of NEM. C) Sample traces show mIPSCs in presence of NEM (left trace) and together with KA (right trace). D, E) A summary bar graph shows that KA has no effect on either mIPSCs frequency or amplitude in the presence of NEM. (See reference 13 for more details.) From X-T Jin, Smith Y et al. Neuroscience 2007; 149:338-349[13] © 2007 with permission from Elsevier.

the secondary activation of G protein-coupled receptors. Thus, together with our recent study showing KAR-mediated regulation of glutamatergic transmission,[9] these findings demonstrate that KARs mediate their presynaptic effects on both GABAergic and glutamatergic transmission in the GP through a metabotropic mode of action.

Substantia Nigra Pars Compacta

Dopaminergic neurons in the substantia nigra pars compacta (SNc) express the highest level of mRNA for GluR5 and GluR7 subunits within the basal ganglia.[16] In contrast, only low level of GluR6 and KA2 is found in SNc neurons, while no detectable mRNA for KA1 is observed in these cells.[16] Although the expression of GluR6, GluR7 and KA2 subunits in SNc was confirmed by light microscopic immunocytochemistry,[41] much remains to be known about the exact cellular and subcellular localization of these subunits in the SNc.

Kainate application increases the frequency of mIPSCs, without changing their amplitude, in SNc neurons.[35] This presynaptic facilitatory effect of KA on the frequency of mIPSCs is suppressed in either Na^+-free or Ca^2-free external solution and in presence of voltage-dependent Ca^{2+} channel

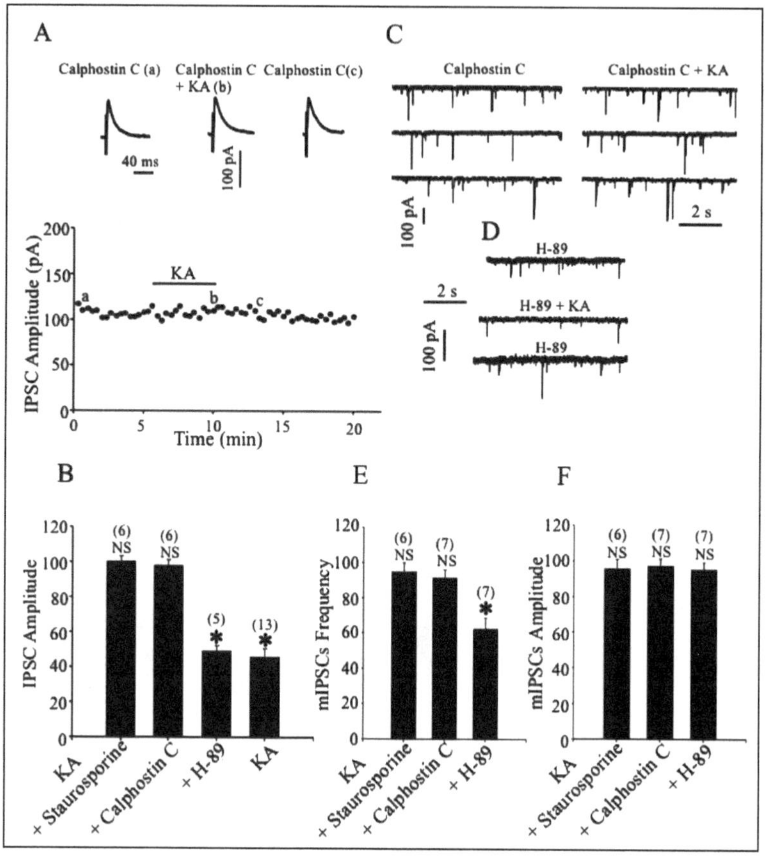

Figure 6. Pretreatment with PKC inhibitor (calphostin), but not PKA inhibitor (H-89), prevents KAR activation-induced inhibition of IPSCs in rat GP. A) The time course of 1 μM KA on IPSC amplitude in the presence of 1 μM calphostin C. Three IPSCs are averaged in each trace at the time indicated by corresponding letters in the graph. B) A bar graph shows that the KAR activation-induced inhibition of IPSCs is blocked by 0.5 μM staurosporine, a broad-spectrum inhibitor of protein kinase and calphostin C, but not by H-89. There is a significant difference from control, * < 0.01. C) mIPSCs were recorded in the presence of 1 μM calphostin C and calphostin C together with 1 μM KA. D) mIPSCs were recorded in the presence of 0.5 μM H-89 together with 1 μM KA. E) Summary bar graph shows that KA has no effect on mIPSCs frequency in the presence of 0.5 μM staurosporine and calphostin C, but not in the presence of H-89. F) Summary bar graph shows that KA has no effect on mIPSCs amplitude in the presence of staurosporine, calphostin C and H-89. *P < 0.05. The y axis of all graphs respresents percent of control. (See ref. 13 for more details.) From X-T Jin, Smith Y et al. Neuroscience 2007; 149:338-349[13] © 2007 with permission from Elsevier.

blockers supporting a direct presynaptic ionotropic mode of action.[35] On the other hand, KAR activation inhibits eIPSCs and this inhibitory effect is reduced in the presence of $GABA_B$ receptor antagonist, but not other G-protein coupled-receptor antagonists (Fig. 7). Several hypotheses have been proposed to explain the possible mechanisms (s) underlying the presynaptic KAR-induced modulation of GABAergic transmission in various brain regions (reviewed by references 7 and 8). These include direct inhibition of GABA release from terminals,[13,31,37,42,43] a modulation of axonal

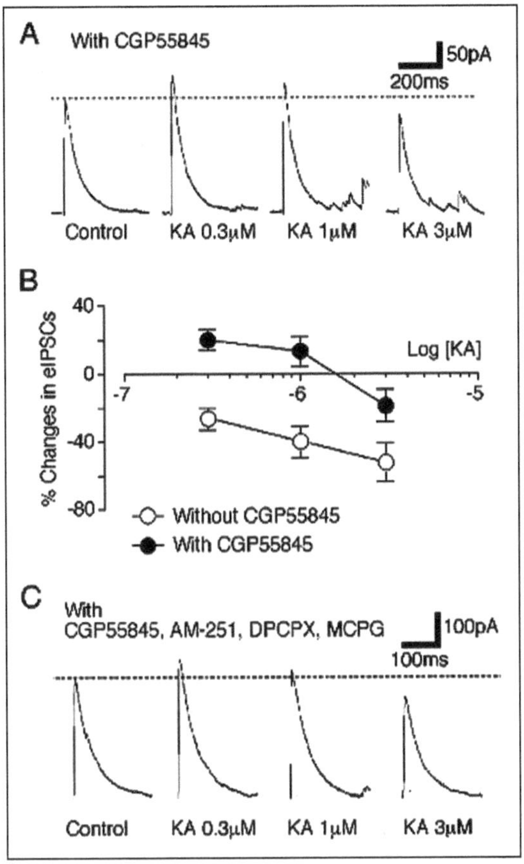

Figure 7. Involvement of GABAB receptors in the KA-induced inhibition of eIPSCs in dopaminergic neurons on the SNc. A) Typical traces of GABAergic eIPSCs observed during application of KA at various concentrations in the presence of 10 μM CGP55845. Dotted line represents control eIPSCs amplitude. B) Concentration-response relationship of KA action on eIPSC amplitude. Each point is normalized to respective control and represents the mean of 6 neurons. C) Typical traces of GABAergic eIPSCs observed during application of KA at various concentration in presence of CGP55845 (10 μM), AM-251 (10 μM), DPCPX (100 nM) and MCPG (1 mM). Reprinted from with permission from Nakamura M et al. J Neurophysiol 2003; 90:1662-1670.[35]

excitability,[44,45] or an indirect $GABA_B$ receptor-mediated effect.[26,27] Data shown in Figure 7 argue that KA-induced inhibition of eIPSCs in the SNc involve the secondary activation of $GABA_B$ autoreceptors. Thus, presynaptic KARs regulate the SNc circuitry through two functionally distinct opposing mechanisms: a direct presynaptic ionotropic mode of action that facilitates GABA release and a secondary inhibitory mechanism that involves presynaptic $GABA_B$ autoreceptors.

In contrast to SNc, neurons in the substantia nigra pars reticulata (SNr) express the highest level of mRNA for GluR5 and GluR6 subunits within the basal ganglia, while no detectable mRNA for the other kainate receptor subunits is found in these neurons.[16] However, immunocytochemical study showed that protein expression of both GluR6/7 and KA2 in the rat SNr.[41] The role of KARs in the SNr remains to be established.

Subthalamic Nucleus

Neurons in subthalamic nucleus (STN) express a high level of mRNA for the GluR6 subunit while they display very low level of GluR7 mRNA, or are almost completely devoid of GluR5, KA1 and KA2 subunits mRNA expression in rats.[16] The high expression level of the GluR6 subunit was confirmed by immunocytochemistry in the rat STN.[41] In contrast to mRNA data, a significant level of KA2 subunit immunoreactivity is also found in the rat STN.[41] The physiology of KARs in STN remains unknown.

Conclusion

The past decade has witnessed significant development in our understanding of KARs-mediated regulation of GABAergic synaptic transmission in the CNS.[7,8,32,43] In the hippocampus, there is significant evidence that converges towards two principal mechanisms of action; a G-protein-coupled, PKC-dependent, metabotropic mechanism or the secondary activation of G-protein coupled receptors.[26,31,36,37] Data reviewed in this chapter provide further evidence for these two pathways in the modulation of GABA release in some basal ganglia nuclei. However, the findings presented in this chapter also highlight a significant degree of heterogeneity by which KARs mediate their effects across basal ganglia nuclei. Future studies aimed at characterizing the localization and function of KARs in the basal ganglia of animal models of Parkinson's disease or other movement disorders may provide some insight about the potential role these receptors may play in the pathophysiology of movement disorders. Knowing that overactive glutamatergic transmission is a cardinal feature of basal ganglia pathophysiology in PD, a deeper understanding of KARs combined with the development of novel compounds that could selectively modulate activity of these receptors may pave the way for new pharmacotherapeutic approaches in PD and other movement disorders.

Acknowledgements

This research was supported by a grant from the US Army, the Yerkes Primate Center NIH base Grant and an Award from Merck/Center for Neurodegenerative Disease at Emory University.

References

1. Hollman M, Heinemann S. Cloned glutamate receptors. Annu Rev Neurosci 1994; 17:31-108.
2. Bettler B, Mulle C. Neurotransmitter receptors II. AMPA and kainat ereceptors. Neuropharmacology 1995; 34:123-139.
3. Charara A, Blankstein E, Smith Y. Presynaptic kainate receptors in the monkey striatum. Neuroscience 1999; 91(4):1195-1200.
4. Lerma J, Paternain AV, Rodriguez-Moreno A et al. Molecular physiology of kainate receptors. Physiol Rev 2001; 81:971-998.
5. Kieval JZ, Hubert GW, Charara A et al. Subcellular and subsynaptic localization of presynaptic and postsynaptic kainate receptor subunits in the monkey striatum. J Neurosci 2001; 21:8746-8757.
6. Kullmann DM. Presynaptic kainate receptors in the hippocampus: slowly emerging from obscurity. Neuron 2001; 32:561-564.
7. Huettner JE. Kainate receptors and synaptic transmission. Prog in Neurobiol 2003; 70(5):387-407.
8. Lerma J. Roles and rules of kainate receptors in synaptic transmission. Nat Rev Neurosci 2003; 4(6):481-495.
9. Jin X-T, Paré JF, Raju DV et al. Localization and function of pre and postsynaptic kainate receptors in the rat globus pallidus. Eur J Neurosci 2006; 23:374-386.
10. Melyan Z, Lancaster B, Wheal HV. Metabotropic regulation of intrinsic excitability by synaptic activation of kainate receptors. J Neurosci 2004; 24:4530-4534.
11. Braga MFM, Aroniadou-Anderjaska V, Xie J et al. Bidirectional modulation of GABA release by presynaptic glutamate receptor 5 kainate receptors in the basolateral amygdala. J Neurosci 2003; 23(2):442-452.
12. Cossart R, Tyzio R, Dinocourt C et al. Presynaptic kainate receptors that enhance the release of GABA on CA1 hippocampal interneurons. Neuron 2001; 29(2):497-508.
13. Jin X-T, Smith Y. Activation of presynaptic kainate receptors suppresses GABAergic synaptic transmission in the globus pallidus. Neuroscience 2007; 149:338-349.
14. Smith Y, Charara A, Paquet M et al. Ionotropic and metabotropic GABA and glutamate receptors in primate basal ganglia. J Chem Neuroanat 2001; 22:13-42.
15. Bahn S, Volk B, Wisden W. Kainate receptor gene expression in the developing rat brain. J Neurosci 1994; 14(9):5525-5547.

16. Bischoff S, Barhanin J, Bettler B et al. Spatial distribution of kainate receptor subunit mRNA in the mouse basal ganglia and ventral mesencephalon. J Comp Neurol 1997; 379:541-562.
17. Chergui K, Bouron A, Normand E et al. Functional GluR6 kainate receptors in the striatum: indirect downregulation of synaptic transmission. J Neurosci 2000; 20(6):2175-2182.
18. Crowder TL, Ariwodola OJ, Weiner JL. Kainate receptor activation potentiates GABAergic synaptic transmission in the nucleus accumbens core. Brain Res 2006; 73-82.
19. Koos T, Tepper J. Inhibittory control of neurostriatal projection neurons by GABAergic interneurons. Nat Neurosci 1999; 2:467-472.
20. Tepper JM, Bolam JP. Functional diversity and specificity of neostriatal interneurons. Curr Opin Neurobio 2004; 14:685-692.
21. Jaeger D, Kita H, Wilson CJ. Surround inhibition among projection neurons is week or nonexistent in the rat neostriatum. J Neurophysio 1994; 72:2555-2558.
22. Kita H, Kitai ST. Intracellular study of rat globus pallidus neurons: membrane properties and responses to neostriatal, subthalamic and nigral stimulation. Brain Res 1991; 564:296-305.
23. Gerfen CR. The neostriatal mosaic: multiple levels of compartmental organization in the basal ganglia. Annu Rev Neurosci 1992; 15:285-320.
24. Bevan MD, Booth PA, Eaton SA et al. Selective innervation of neostriatal interneurons by a subclass of neuron in the globus pallidus of the rat. J Neurosci 1998; 18:9438-9452.
25. Smith Y, Bevan MD, Shink E et al. Microcircuitry of the direct and indirect pathways of the basal ganglia. Neuroscience 1998; 86:353-387.
26. Frerking M, Petersen CC, Nicoll RA. Mechanisms underlying kainate receptor-mediated disinhibition in the hippocampus. Proc Natl Acad Sci USA 1999; 96:12917-12922.
27. Kerchner GA, Wilding TJ, Li P et al. Presynaptic kainate receptors regulate spinal sensory transmission. J Neurosci 2001; 21:59-66.
28. Hettinger BD, Lee A, Linden J et al. Ultrastructural localization of adenosine A_{2A} receptors suggests multiple cellular sites for modulation of GABAergic neurons in rat striatum. J Comp Neurol 2001; 431:331-346.
29. Bogenpohl JW, Pare JF, Smith Y. Subcellular localization of adenosine A_{2A} receptors in the striatum and globus pallidus of monkey and rat. Soc Neurosci Abstr 2008; P-056.
30. Kane-Jackson R, Smith Y. Pre-synaptic kainate receptors in GABAergic and Glutamatergic axon terminals in the monkey globus pallidus. Neuroscience 2003; 120:285-289.
31. Rodríguez-Moreno A, Lerma J. Kainate receptor modulation of GABA release involves a metabotropic function. Neuron 1998; 20:1211-1218.
32. Frerking M, Nicoll RA. Synaptic kainate receptors. Curr Opin Neurobiol 2000; 10:342-351.
33. Rodríguez-Moreno A, Sihra TS. Metabotropic actions of kainate receptors in the CNS. J Neurochem 2007; 103:2121-2135.
34. Rodríguez-Moreno A, Sihra TS. Kainate receptors with a metabotropic modus operandi. Trends Neurosci 2007; 30(12):630-637.
35. Nakamura M, Jang IS, Ishibashi H et al. Possible roles of kainate receptors on GABAergic nerve terminals projecting to rat substantia nigra dopaminergic neurons. J Neurophysiol 2003; 90:1662-1670.
36. Cunha RA, Malva JO, Ribeiro JA. Pertussis toxin prevents presynaptic inhibition by kainate receptors of rat hippocampal [^{3}H] GABA release. FEBS Letters 2000; 469(2-3):159-162.
37. Rodriguez-Moreno A, Lopez-Garcia JC, Lerma J. Two populations of kainate receptors with separate signaling mechanisms in hippocampal interneurons. Proc Natl Acad Sci USA 2000; 97:1293-1298.
38. Melyan Z, Wheal HV, Lancaster B. Metabotropic-mediated kainate receptor regulation of IsAHP and excitability in pyramidal cells. Neuron 2002; 34(1):107-114.
39. Melyan Z, Lancaster B, Wheal HV. Metabotropic regulation of intrinsic excitability by synaptic activation of kainate receptors. J Neurosci 2004; 24:4530-4534.
40. Lauri SE, Segerstrale M, Vesikansa A et al. Endogenous activation of kainate receptors regulates glutamate release and network activity in the developing hippocampus. J Neurosci 2005; 25:4473-4484.
41. Petralia RS, Wang YX, Wenthold RJ. Histological and ultrastructural localization of the kainat receptor subunits, KA2 and GluR6/7, in the rat nervous system using selective antipeptide antibodies. J Comp Neurol 1994; 349:85-110.
42. Rodriguez-Moreno A, Herreras O, Lerma J. Kainate receptors presynaptically downregulate GABAergic inhibition in the rat hippocampus. Neuron 1997; 19:893-901.
43. Maingret F, Lauri SE, Taira T et al. Profound regulation of neonatal CA1 rat hippocampal GABAergic transmission by functionally distinct kainate receptor populations. J Physiol 2005; 567:131-142.
44. Semyanov A, Kullmann DM. Kainate receptor-dependent axonal depolarization and action potential initiation in interneurons. Nature Neurosci 2001; 4:718-723.
45. Kang N, Jiang L, He W et al. Presynaptic inactivation of action potentials and postsynaptic inhibition of GABAA currents contribute to KA-induced disinhibition in CA1 pyramidal neurons. J Neurophysiol 2004; 92:873-882.

Chapter 4

Metabotropic Actions of Kainate Receptors in the Control of Glutamate Release in the Hippocampus

Antonio Rodríguez-Moreno* and Talvinder S. Sihra*

Abstract

Kainate-type glutamate receptors (KARs) structurally present the credentials of the other ionotropic glutamate receptor (iGluR) family members (NMDA and AMPA receptors), but functionally often purport examples of a metabotropic mode of operation. In the present chapter, we describe these metabotropic roles of KARs in the modulation of glutamate release in the hippocampus at CA3 Schaffer Collateral (SC)-CA1 Pyramidal Cell (PC) synapses and dentate gyrus granule cell Mossy Fiber (MF)-CA3 PC synapses. As autoreceptors on SC terminals, KARs inhibit the release of glutamate at SC-CA1 PC synapses through a mechanism dependent on a pertussis toxin-sensitive $G_{i/o}$ protein thought to couple via its $G\beta\gamma$ subunit to a decrease in Ca^{2+} channel function. At MF-CA3 PC synapses, autoreceptors on MF terminals respond diametrically depending on the agonist concentration. At low KA concentrations (<100 nM), a G-protein-independent process invokes the activation of proteins kinase A (PKA) to effect a facilitation of glutamate release. This facilitation possibly involves the Ca^{2+}-dependent (rather than GPCR-dependent) activation of adenylate cyclase (AC). At high KA concentrations (>100 nM), a mechanism involving a pertussis toxin-sensitive $G_{i/o}$ protein is invoked to inhibit AC activity and thereby suppress PKA activity. Taken together with the heterosynaptic regulation of GABA release by KARs working with a metabotropic *modus operandi*, there is therefore compelling evidence that these ionotropic glutamate receptors are involved in a noncanonical modulation of glutamate release that does not rely on their typical ionotropic activity.

Introduction

Glutamate is the major excitatory neurotransmitter in the mammalian central nervous system and as such participates in normal synaptic transmission, as well as processes thought to underlie learning and memory, including long-term potentiation (LTP) and long-term depression (LTD).[1] Developmentally, glutamate plays essential roles during neuronal maturation and synaptogenesis. Alterations in the glutamate system may contribute to the aetiology of a number of neuropathologies, including neuronal degeneration arising following brain ischaemia and hypoglycaemia, Alzheimer's, Parkinson's, Huntington chorea, lateral amyotrophic sclerosis and some forms of epilepsy.[1-6] Clearly, the modulation of glutamate release is an essential requisite to maintain the

*Corresponding Authors: Antonio Rodríguez-Moreno—Departamento de Fisiología, Anatomía y Biología Celular, Universidad Pablo de Olavide, Ctra. de Utrera, Km. 1, 41013, Sevilla Spain. Email: arodmor@upo.es
Talvinder S. Sihra— Department of Neuroscience, Physiology and Pharmacology, University College London,Gower Street, London, UK. Email: t.sihra@ucl.ac.uk

Kainate Receptors: Novel Signaling Insights, edited by Antonio Rodríguez-Moreno and Talvinder S. Sihra. ©2011 Landes Bioscience and Springer Science+Business Media.

high fidelity of synaptic transmission. Presynaptic metabotropic glutamate receptors (mGluRs) offer one layer of the autoreceptor control of glutamate release, but, as with the control of GABA, ionotropic glutamate receptors (iGluRs) also afford regulatory control presynaptically. Of the three iGluR subtypes, viz NMDA, AMPA and Kainate receptors, the kainate receptor (KAR) has long been thought to be active presynaptically based on studies showing glutamatergic lesions consequent from the neurotoxic consequences of kainate (KA) administration.[7]

In order to address the regulatory influences of KARs, the hippocampus, with its well defined synaptic fields, represents an excellent model. In the hippocampus, the major projecting pathways are glutamatergic. The glutamatergic output from dentate gyrus granule cells forms mossy fibre (MF)-CA3 pyramidal cell (PC) synapses, while axons from the CA3 PCs project to the CA1 field to form Schaffer Collateral (SC)-CA1 pyramidal cell (PC) synapses, as well providing input to interneurons for disynaptic inhibition of CA1 PCs.

KARs on Schaffer Collateral (SC) Terminals Suppress Glutamate Release at Schaffer Collateral (SC)-CA1 Pyramidal Cell (PC)

At the CA3 Schaffer Collateral (SC)-CA1 pyramidal cell (PC) synapse, KAR activation by KA, or the low affinity KAR agonist domoic acid (DA), effects a suppression of the evoked excitatory postsynaptic potential eEPSP measured in CA1 PCs, suggesting that KAR activation inhibits glutamate release[8-13] (Fig. $1b_1$). This effect is pharmacological consistent with the exclusive involvement of KARs.[11] Moreover, the presynaptic localization of these KARs is confirmed based on observations that KA invokes an increase in paired pulse facilitation and causes a decrease in $1/CV^2$ that correlates with a decrease in the mean EPSC amplitude, both parameter being diagnostics of a presynaptic locus of receptor action.[11]

A priori, mechanistically, the KA-induced modulation of glutamate release could be attributed to an ionotropic action of presynaptic KARs, whereby depolarization leads to the voltage-dependent inactivation of the Ca^{2+} channels supporting glutamate release.[8,9] However, Frerking et al[11] argue that presynaptic KAR activation produces a decrease of glutamate release through a metabotropic mechanism of action. This contention is based on experiments showing that the reduction in the eEPSP produced by KAR activation by DA, is abrogated by the G-protein inhibitors, N-ethylmaleimide (NEM) and pertussis toxin (PTX) (Fig. $1a_1$ and b_1). This demonstrates the overt involvement of $G_{i/o}$ in the inhibition of glutamate release by KARs. Intriguingly though, the modulation is not affected by the broad spectrum protein kinase inhibitor H-7 (Fig. $1a_2$ and b_2), indicating that 2nd messenger-mediated stimulation of protein kinases does not play a role.[11] Rather, the regulation may be intrinsic to plasma membrane, whereby membrane-delimited $\beta\gamma$ subunits of $G_{i/o}$ directly bind to and inhibit presynaptic Ca^{2+} channels to thus suppress glutamate release,[11] similarly to the action of several inhibitory presynaptic GPCRs[14] (Fig. 1c).

Consistent with a KAR-mediated reduction of presynaptic Ca^{2+} channel activity in SC terminals, Kamiya and Ozawa[9] have shown that the activation of KARs produces a decrease in intracellular Ca^{2+} concentration that correlates with the inhibition of glutamate release. The decrease in Ca^{2+} channel activity produced by KAR is not a consequence from some reduced excitability of the nerve terminal as there is no change in the presynaptic (afferent) fiber volley upon application of KA. This accentuates the lack of involvement of any ionotropic, depolarizing influences of KAR activation in the modulation observed. Although KA application does produce an inward current while decreasing glutamate release in slices,[8] the inward current recovers rapidly and completely, this being in marked contrast to long-lasting effect of KA on the eEPSC. Altogether, these observations point to Schaffer collateral nerve terminals expressing autoregulatory KARs which depress glutamate release through a metabotropic mode of operation, likely involving a reduction in voltage-dependent Ca^{2+} channel activity.

Support for the metabotropic operation of KARs in suppressing glutamate release has come from studies looking at the CA3-CA1 PC synapses during development.[13] Additionally, apart from the CA3 Schaffer collateral input to CA1 PCs, CA3 PCs also project association/commissural (A/C) fibres which have terminals synapsing onto the apical dendrites of CA3 PC in the

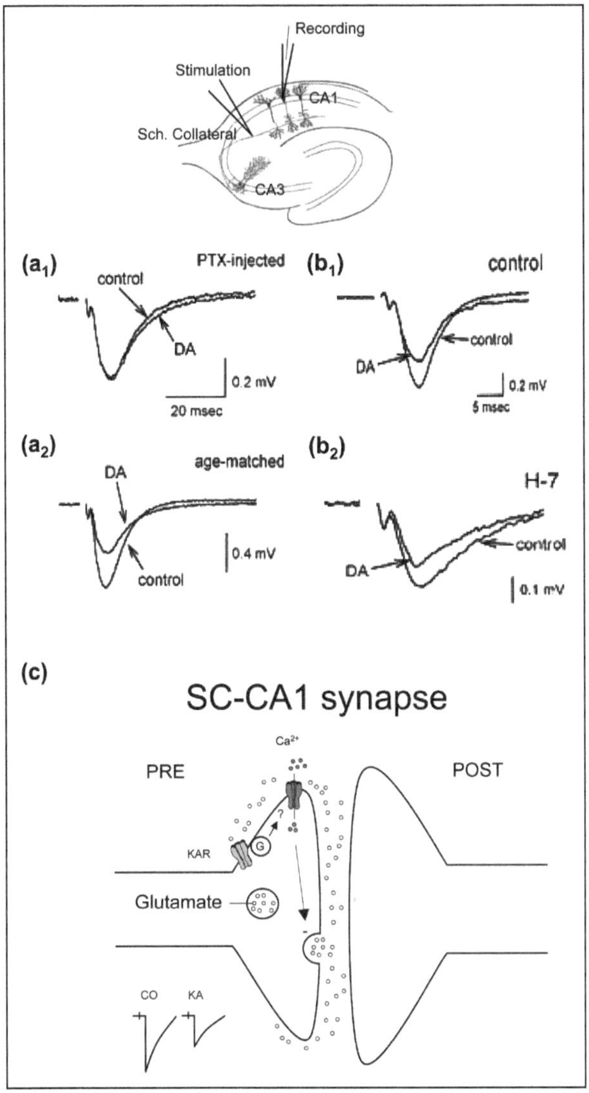

Figure 1. Metabotropic actions of KARs in the regulation of glutamate release at SC-CA1 synapses in hippocampal slices. Inset, experimental setup. $a_{1,2}$) Domoate (DA), a low affinity KAR-agonist mediates a depression of fEPSP amplitudes at SC-CA1 synapses which is abolished in slices treated with pertussis-toxin (PTX). $b_{1,2}$) The depression of the fEPSP amplitude is evidently not dependent on any protein kinase cascades as a broad spectrum protein kinase inhibitor H7 does not prevent the DA action. c) Schematic of signalling at the SC-CA1 synapse. Metabotropic actions of KARs at Schaffer collaterals (SC)-CA1 pyramidal cell synapses. Activation of presynaptic KARs on glutamatergic SC terminals (PRE) projecting from CA3 pyramidal neurons decreases glutamate release onto CA1 neuron dendrites in the *stratum radiatum* (POST). This modulation involves presynaptic G-protein activation and regulation of voltage-dependent Ca^{2+} entry. Panel c modified and reproduced from Rodríguez-Moreno, Sihra TS. Trends Neurosci 2007; 30:630-637,[44] ©2007 with permission from Elsevier.

stratum radiatum. Glutamate released from these terminals can heterosynaptically modulate the major MF-CA3 PC synapse in the adjacent *stratum lucidum*. Notably, the A/C terminals also express KARs that inhibit glutamate release. Given the common origin of SC and A/C terminals projecting from CA3 PCs, it is plausible that these terminals might share a common metabotropic mechanism of KAR-mediated inhibition of glutamate release, although this remains to be established experimentally.

KARs on Granule Cells (Dentate Gyrus) Mossy Fibre (MF) Terminals Effect Bimodal Modulation of Glutamate Release at Mossy Fibre (MF)-CA3 Pyramidal Cell (PC) Synapses

Mossy fibre (MF)-CA3 pyramidal cell (PC) synapses have long been known to display frequency facilitation.[15] One possibly means of achieving this type of synaptic plasticity would be iGluRs on MF terminals operating as facilitatory autoreceptors. Consistent with the operation of facilitatory KARs, KA application causes a potentiation of secretagogue, 4-aminopyridine (4-AP)-evoked release of glutamate release from hippocampal synaptosomes (Fig. 2a,b). In slice experiments measuring glutamate release at the intact MF-CA3 PC synapse, low concentrations of KA (30 nM) increase the evoked excitatory postsynaptic current (eEPSC) recorded in CA3 PC following electrical stimulation of MFs (Fig. 2c,d,e). This suggests that synaptically released glutamate could well be activating KARs located at the MF terminals, homo- and hetero-synaptically and thereby instigates the frequency-facilitation seen at the MF-CA3 PC synapse.

The mechanism of KA-mediated facilitation of glutamate release in general clearly remains contentious. A parsimonious explanation for the facilitation could attribute the modulation to a classical ionotropic or depolarizing effect of KAR activation. However, in a study by Perkinton and Sihra[16] using synaptosomes, although KA enhanced 4-AP-induced glutamate release, secretagogue-induced changes in ionotropic parameters such as membrane potential or intracellular $[Ca^{2+}]$, were not affected by the addition of KA. Although it is possible that the averaging effect of measurements from a synaptosomal population might mask a small local effect on a subset of nerve terminals expressing KARs, the data do not generally support an ionotropic mechanism for the facilitation of glutamate observed. Actually, given the strong stimulation that might be expected at the KA concentrations used,[17,18] an ionotropic effect of KARs might well inactivate voltage-dependent Na^{+}- and/or Ca^{2+}-channels and thereby inhibit 4-AP-mediated stimulus-release coupling. This is not observed for glutamate release in this preparation, though at the same time, GABA release is seen to be inhibited by KA treatment (see previous chapter herein).

Notwithstanding the data from isolated nerve terminals, in electrophysiological studies using hippocampal slices, given that low concentrations of KA cause an after-depolarization of MF terminals, there is the possibility that this underpins an increase in nerve terminal Ca^{2+} levels. As such, this would thereby provide for an ionotropic mechanism for the enhancement of glutamate release leading to frequency facilitation seen at the MF-CA3 PC synapse. Although these data are in line with an ionotropic mechanism for the KAR modulation, curiously the aforementioned after-depolarization of MF terminals is somewhat long-lasting, begging the question: is there is metabotropic influence of KAR at MF-CA3 PC synapses too?

Consistent with this notion, potentiation of glutamate release from hippocampal synaptosomes is inhibited by H-89, a catalytic inhibitor of cAMP-dependent protein kinase (PKA) (Fig. 2a,b). Moreover, application of forskolin + IBMX, a treatment known to elevate synaptosomal cAMP levels and thereby PKA activity, occludes the facilitatory effect of KA (Fig. 2a,b). In the hippocampal slice preparation, the KA-mediated enhancement of the eEPSC is suppressed by H-89, as well as Rp-Br-cAMP, which is inhibits the activation of PKA by endogenous cAMP (Fig. 2c,e). Elevation of cAMP in slices using the adenylate cyclase (AC) activator, forskolin, in combination with the phosphodiesterase inhibitor, IBMX, causes synaptic potentiation of the MF-CA3 PC synapse, but under these conditions, the facilitatory effect of KA is occluded (Fig. 2c,e). Evidently, the facilitation of glutamate release seen with low concentrations of KA involves an adenylate cyclase/cAMP/protein kinase A (AC/cAMP/PKA) signalling cascade, such that

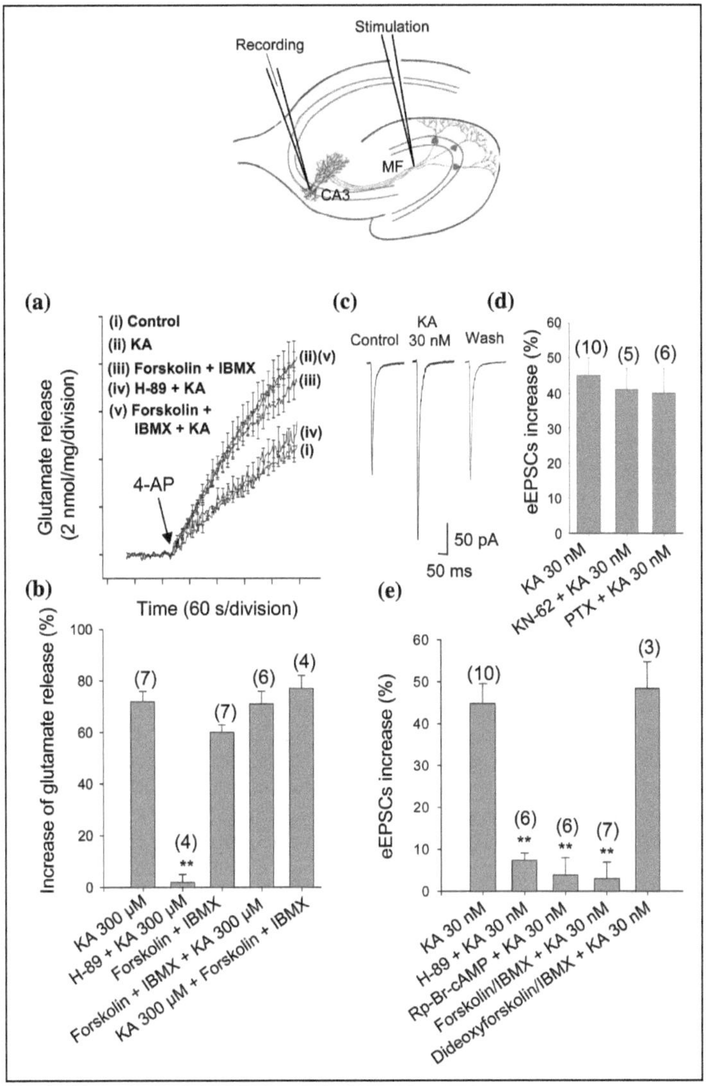

Figure 2. KARs activation by low agonist concentration produces an increase of glutamate release that involves the activation of PKA at MF-CA3 synapses. Inset, experimental setup. a) Activation of adenylyl cyclase and downstream protein kinase A underlies KAR-mediated facilitation of 4-aminopyridine (4-AP)-evoked glutamate release from hippocampal synaptosomes. Glutamate release under control conditions (i) and in the presence of (ii) KA, (iii) forskolin/IBMX (iv) H-89 + KA (iv) and (v) forskolin + IBMX + KA. b) Quantification of KA effects described in (a). c) Low concentrations of KA (30 nM) increase the eEPSC amplitude in hippocampal slices in the presence of AP-5 and bicuculline. d) The KA-induced increase in glutamate release was unaltered by treating the slices with the CaMKII inhibitor, KN62, or with the G-protein inhibitor pertussis toxin (PTX), but is prevented in the presence of H-89 and Rp-BrcAMP and occluded by previous addition of forskolin (e). Adapted from Rodríguez-Moreno A, Sihra TS. J Physiol 2004; 557:733-745,[19] ©2004 with permission from Wiley-Balckwell.

prior activation of the signalling pathway mitigates the facilitatory effects of KARs at MF-CA3 synapses.[19] Perhaps unsurprisingly, the facilitatory presynaptic effects of KA on MF-CA3 synapses (Fig. 2c) are not however prevented by the inhibition of $G_{i/o}$ by PTX, confirming that KAR receptor activity impinges at, or downstream of, AC in the AC/cAMP/PKA signalling cascade.

How does the ionotropic and metabotropic evidence for the facilitatory function of KARs on glutamate release mutually consolidate? One possible sequence of events that might produce a facilitation of glutamate release, may first be contingent on an increase in intraterminal [Ca^{2+}] produced by KAR activation. Given this, a Ca^{2+}-dependent activation of the AC/cAMP/PKA pathway is plausible if small local increases in intracellular Ca^{2+} stimulate Ca^{2+}/CAM-dependent ACs. This hypothesis was first postulated in relation to the synaptic plasticity observed at the MF-CA3 synapse,[20] although the involvement of Ca^{2+}/CAM kinase II invoked therein, does not apply to KAR activity as the Ca^{2+}/CAM kinase II inhibitor, KN-62 had no effect on the increase in eEPSCs induced by KA. Interestingly however, recent work indicates that KARs can gate extracellular Ca^{2+}-influx to produce a Ca^{2+}-induced intracellular Ca^{2+} release which supports facilitation and LTP (long-term potentiation) at MF-CA3 PC synapses.[21] Whether this type of intracellular Ca^{2+} release subserves the activation of an AC/cAMP/PKA pathway to effect the potentiation, remains to be seen.

An intriguing aspect of KAR function at the MF-CA3 PC synapse is that, unlike the monotonic inhibitory regulation by KARs at SC-CA1 PC synapses, this synapse exhibits bidirectional control of glutamate release. Thus while, nanomolar concentrations of KA (<50 nM) facilitate glutamate release (Fig. 2),[19,21-26] higher concentrations agonist (>100 nM) inhibit glutamate release (Fig. 3a, right).[27-32] The inhibitory KARs are certainly presynaptic given that KA application alters paired pulse facilitation, increases the number of failures of eEPSC and effects a change in the coefficient of variation ($1/CV^2$) that correlates with the change in synaptic response.[31] The question is: how does KAR activity enhance or reduce glutamate release from the same synapses, or indeed the same terminals, depending on agonist concentration? Again the ionotropic argument could follow: low concentrations of KA cause depolarisation and thereby inactivate K^+-channels and/or increase Ca^{2+} influx to enhance release. Higher concentrations of KA effect stronger depolarisation and thus cause inactivation of Na^+ and/or Ca^{2+} channels and/or electrical shunting, to thereby reduce terminal excitability and hence decrease glutamate release.[27,29,33] Like the KAR-mediated facilitation of glutamate release discussed above, notwithstanding the potential ionotropic consequences of KAR activation of MF-CA3 PC synapses by high KA concentrations, closer examination of the KAR/glutamate release suppression profile reveals key metabotropic features (Fig. 3).[31]

Firstly, alluding to a metabotropic coupling of inhibitory KAR/glutamate release coupling, the effect of high [KA] application at MF-CA3 PC synapses is found to be long-lasting and with a slow recovery. This is at variance with the proposed ionotropic mechanism of KAR-mediated modulation at MF-CA3 synapses, which invokes that the change and recovery of the KA-evoked holding current is fast.[27] Notwithstanding the temporal arguments, pharmacological data further support the metabotropic nature of the KAR-mediated depression of glutamate release at MF-CA3 PC synapses. Crucially in this regard, treatment of the slices with the $G_{i/o}$ inhibitor PTX prevents the depression of glutamate release (Fig. 3a, left). Together with this implication of $G_{i/o}$ involvement in the KAR-mediated modulation, inhibition of PKA with the catalytic PKA inhibitor, H-89, or the cAMP competitor Rp-Br-cAMP, occludes the inhibitory effect of KA on the EPSC, the inhibitors themselves having already suppressed the EPSC (Fig. 3c, d). The indication from this is that high [KA] evoked KAR activation elicits a $G_{i/o}$-mediated reduction in the AC/cAMP/PKA signalling cascade to effect a suppression of glutamate release. Consistent with this, when cAMP and thus PKA activity, is "clamped" high by using an exogenous cAMP analogue, Sp-8-CPT-cAMPs, KA is unable to suppress the ESPC.[31] Interestingly, although the inhibition of glutamate release seen here at the MF-CA3 PC synapse resembles the inhibition of GABA release at the interneuron-CA1 PC synapse in being dependent on $G_{i/o}$, they differ in the downstream signalling. Thus while PKA *inhibition* underpins the KAR-mediated suppression of the EPSC decrease, PKC *activation* (calphostin C-sensitive) supports the KAR-mediated suppression of the eIPSC. Note that calphostin C, has no effect on KAR function at MF-CA3 PC synapses (Fig. 3d, cf KAR-function at interneuron-CA1 PC synapses; see previous chapter herein).

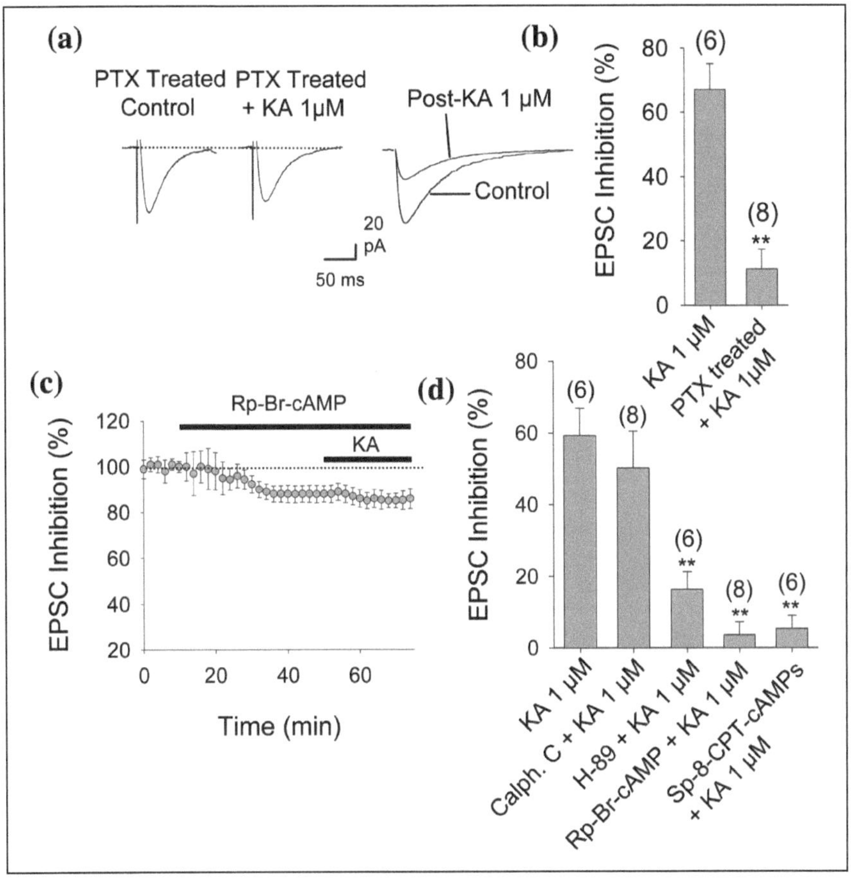

Figure 3. Metabotropic action of KARs depressing glutamate release at MF-CA3 synapse. a) KA depresses glutamate release at MF-CA3 synapses, an effect that is prevented in the presence of pertussis toxin (a,b) and PKA inhibition (Rp-Br-cAMP and H-89), but not by PKC inhibition (calphostin C, Calph. C) (c,d). The effect of KA is occluded by the cAMP analogue, Sp-8-CPT-cAMPs (d). Panels (a-d), adapted from Negrete-Díaz JV et al. J Neurophysiol 2006; 96:1829-1837.[31]

At MF-CA3 PC synapses, the general working hypothesis arising is that increased presynaptic AC/cAMP/PKA signalling facilitates glutamate release while decreased AC/cAMP/PKA signalling suppresses release. Low KA concentrations instigate the former while high KA concentrations invoke the latter. Intriguingly, the decreased AC/cAMP/PKA signalling linked to the decrease of glutamate release at MF-CA3 PC synapses concords with the signalling underpinning the induction of KAR-mediated synaptic depression and low-frequency stimulation (LFS)-mediated long-term depression (LTD).[32] LTD at MF-CA3 synapses has classically been shown to be mediated by type II mGluRs which also decrease the activity of the AC/cAMP/PKA cascade through $G_{i/o}$ activation.[34,35] Thus, two types of glutamate receptors, viz. KAR (in a metabotropic guise) and Type II mGluRs, appear to collude in this form of plasticity and indeed mutually occlude each other when applied consecutively. Neatly, this mirrors the observations with long-term potentiation (LTP) at MF-CA3 PC synapses, where KAR-mediated facilitation and excitatory (Type I), mGluR activation both lead to an increase in glutamate release.[23]

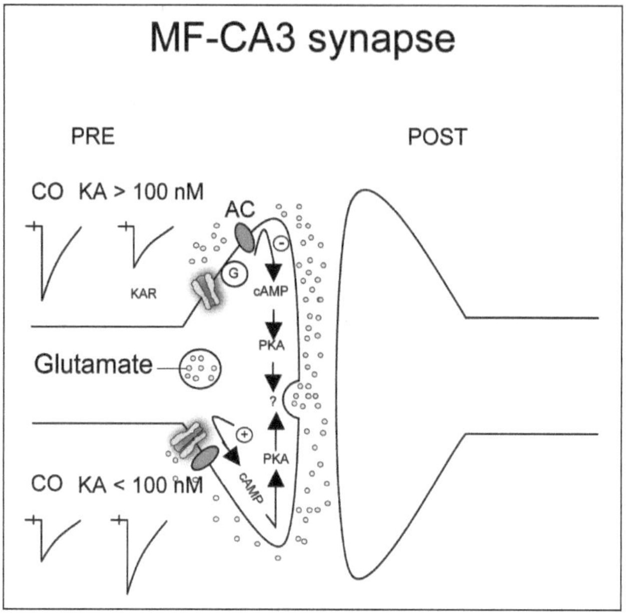

Figure 4. Metabotropic actions of KARs at mossy fiber (MF)-CA3 pyramidal cell synapses. KARs produce a bimodal effect on release from MFs depending on the agonist concentration (PRE): [KA] > 100 nM decrease glutamate release following activation of a G protein and the modulation of adenylate cyclase (AC) and PKA activity. [KA] < 100 nM facilitates glutamate release following activation of AC and PKA. Reproduced from Rodríguez-Moreno, Sihra TS. Trends Neurosci 2007; 30:630-637,[44] ©2007 with permission from Elsevier.

Overall, KARs evidently participate intimately in all the key forms of synaptic plasticity displayed at the MF-CA3 synapse.[32,36] Interestingly, in the developing hippocampus, KARs exhibit metabotropic actions that mediate the regulation of glutamate release and network activity in response to synaptic activation.[37] During the first postnatal week of hippocampal development, endogenous glutamate appears to regulate release in an action potential-independent manner, by tonically activating KARs at CA3 glutamatergic synapses.

Here we have largely concentrated on discussing metabotropic regulation of glutamate release by KARs which is currently matter of active debate. There are indeed instances of KAR-mediated regulation of glutamate release where the modulation is patently ionotropic, or mechanistic details are still under investigation. These include examples in the developing cortex[38] (inhibitory effect), amygdala[39] (facilitatory effect), nucleus accumbens[40,41] (inhibitory effect), dorsal root ganglion cells[42] (inhibitory effect) and cerebellum[43] (facilitatory and inhibitory effect). Presynaptic KARs evidently operate through a combination of ionotropic and metabotropic mechanisms to modulate excitatory glutamatergic transmission.

Conclusion

In summary,[44,45] (Fig. 4) at SC-CA1 synapses, KARs inhibit glutamate release via a metabotropic, G-protein dependent mechanism but protein kinase-independent mechanism. At MF-CA3 synapses there is a biphasic effect, with low concentrations of KA inducing a facilitation of glutamate release mediated by an AC/cAMP/PKA pathway with, no necessity of G-protein and higher KA concentrations producing a depression of glutamate release mediated by an AC/cAMP/PKA pathway and with the necessary participation of a G protein.

Acknowledgements

Work performed by the authors concerning metabotropic actions of KARs in MF-CA3 was supported by Grant BFU2006-14155 from the Spanish MEC (ARM) and a FEBS Short-term Fellowship to AR-M, held in the laboratory of TSS.

References

1. Jonas P, Monyer H. Ionotropic glutamate receptors in the CNS. Springer, Berlin, 1999.
2. Sander T, Janz D, Ramel C et al. Refinement of map position of the human GluR6 kainate receptor gene (GRIK2) and lack of association and linkage with idiopathic and generalized epilepsies. Neurology 1995; 45:1713-1720.
3. Sander T, Hildmann T, Kretz R et al. Allelic association of juvenile absence epilepsy with a GluR5 kainate receptor gene (GRIK1) polymorphism. Am J Med Genet 1997; 74:416-421.
4. Mulle C, Sailer A, Perez-Otano I et al. Altered synaptic physiology and reduced susceptibility to kainate-induced seizures in GluR6-deficient mice. Nature 1998; 392:601-605.
5. McDonald ME, Vonsattel JP, Shrinidi J et al. Evidence for the GluR6 gene associated with younger onset age of Huntington's disease. Neurology 1999; 53:1330-1332.
6. Eubanks JH, Puraman RS, Klecner NW et al. The gene encoding the glutamate receptor subunit GluR5 is located on human chromosome 21q21.1-221 in the vicinity of the gene for familial amyotrophic lateral sclerosis. Proc Natl Acad Sci USA 1993; 90:178-182.
7. Ferkany JW, Zaczek R, Coyle JT. The mechanism of kainic acid neurotoxicity. Nature 1984; 308:561-562.
8. Chittajallu R, Vignes M, Dev KK et al. Regulation of glutamate release by presynaptic kainate receptors in the hippocampus. Nature 1996; 379:78-81.
9. Kamiya H, Ozawa S. Kainate receptor-mediated inhibition of presynaptic Ca^{2+} influx and EPSP in area CA1 of the rat hippocampus. J Physiol 1998; 509:833-845.
10. Vignes M, Clarke VJR, Parry MJ et al. The GluR5 subtype of kainate receptor regulates excitatory synaptic transmission in areas CA1 and CA3 of the rat hippocampus. Neuropharmacology 1998; 37:1269-1277.
11. Frerking M, Schmitz D, Zhou Q et al. Kainate receptors depress excitatory synaptic transmission at CA3-CA1 synapses in the hippocampus via a direct presynaptic action. J Neurosci 2001; 21:2958-2966.
12. Clarke VRJ, Collingridge GL. Characterisation of the effects of ATPA, a GLU_{K5} receptor selective agonist, on excitatory synaptic transmission in area CA1 of rat hippocampal slices. Neuropharmacology 2002; 42:889-902.
13. Lauri SE, Vesikansa A, Segerstrale M et al. Functional maturation of CA1 synapses involves activity-dependent loss of kainate receptor-mediated inhibition of glutamate release. Neuron 2006; 50:415-429.
14. Brown DA, Sihra TS. Presynaptic signaling by heterotrimeric G-proteins in Pharmacology of Neurotransmitter Release (Südhof TC, Starke K, ed.), Handb Exp Pharmacol. Series (Springer) 2008; (184):207-260.
15. Nicoll RA, Schmitz D. Synaptic plasticity at hippocampal mossy fibre synapses. Nat Rev Neurosci 2005; 6:863-76.
16. Perkinton MS, Sihra TS. A high-affinity presynaptic kainate-type glutamate receptor facilitates glutamate exocytosis from cerebral cortex nerve terminals (synaptosomes). Neuroscience 1999; 90:1281-1292.
17. Kullmann DM. Presynaptic kainate receptors in the hippocampus: slowly emerging from obscurity. Neuron 2001; 32:561-564.
18. Semyanov A, Kullmann DM. Kainate receptor-dependent axonal depolarization and action potential initiation in interneurons. Nat Neurosci 2001; 4:718-723.
19. Rodríguez-Moreno A, Sihra TS. Presynaptic kainate receptor facilitation of glutamate release involves protein kinase A in the rat hippocampus. J Physiol 2004; 557:733-745.
20. Weisskopf MG, Castillo PE, Zalutsky RA et al. Mediation of hippocampal mossy fiber long-term potentiation by cyclic AMP. Science 1994; 265:1878-1882.
21. Lauri SE, Bortolotto ZA, Nistico R et al. A role of Ca^{2+} stores in kainate receptor-dependent synaptic facilitation and LTP at mossy fiber synapses in the hippocampus. Neuron 2003; 39:327-341.
22. Contractor A, Swanson GT, Sailer A et al. Identification of the kainate receptors subunit underlying modulation of excitatory synaptic transmission in the CA3 region of the hippocampus. J Neurosci 2000; 20:8269-8278.
23. Schmitz D, Mellor J, Nicoll RA. Presynaptic kainate receptor mediation of frequency facilitation at hippocampal mossy fiber synapses. Science 2001; 291:1972-1976.
24. Lauri SE, Bortolotto ZA, Bleakman D et al. A critical role of a facilitary presynaptic kainate receptor in mossy-fiber LTP. Neuron 2001; 32:697-709.

25. Lauri SE, Delany C, Clarke VEJ et al. Synaptic activation of a presynaptic kainate receptor facilitates AMPA receptor-mediated synaptic transmission at hippocampal mossy fibre synapses. Neuropharmacology 2001; 41:907-915.
26. Breustedt J, Schmitz D. Assessing the role of GLUK5 and GLUK6 at hippocampal mossy fiber synapses. J Neurosci 2004; 24:10 093-10 098.
27. Kamiya H, Ozawa S. Kainate receptor-mediated presynaptic inhibition at the mouse hippocampal mossy fibre synapse. J Physiol 2000; 523:653-665.
28. Schmitz D, Frerking M, Nicoll RA. Synaptic activation of presynaptic kainate receptors on hippocampal mossy fiber synapses. Neuron 2000; 27:327-338.
29. Contractor A, Swanson G, Heinemann SF. Kainate receptors are involved in short- and long-term plasticity at mossy fiber synapses in the hippocampus. Neuron 2001; 29:209-216.
30. Contractor A, Sailer A, Darstein M et al. Loss of kainate receptor-mediated heterosynaptic facilitation at mossy-fiber synapses in $KA2^{-/-}$ mice. J Neurosci 2003; 23:422-429.
31. Negrete-Díaz JV, Sihra TS, Delgado-García JM et al. Kainate receptor-mediated inhibition of glutamate release involves protein kinase A in the mouse hippocampus. J Neurophysiol 2006; 96:1829-1837.
32. Negrete-Díaz JV, Sihra TS, Delgado-García JM et al. Kainate receptor-mediated presynaptic inhibition converges with presynaptic inhibition mediated by Group II mGluRs and long-term depression at the hippocampal mossy fiber-CA3 synapse. J Neural Transm 2007; 114:1425-1431.
33. Kamiya H, Umeda K, Ozawa S et al. Presynaptic Ca^{2+} entry is unchanged during mossy fiber long-term potentiation. J Neurosci 2002; 22:10 524-10 528.
34. Kobayashi K, Manabe T, Takahashi T. Presynaptic long-term depression at the hippocampal mossy fiber-CA3 synapse. Science 1996; 273:648-650.
35. Tzounopoulos T, Janz R, Sudhof TC et al. A role for cAMP in long-term depression at hippocampal mossy fiber synapses. Neuron 1998; 21:837-845.
36. Bortolotto ZA, Nistico R, More JC et al. Kainate receptors and mossy fiber LTP. Neurotoxicology 2005; 26:769-777.
37. Lauri SE, Segerstale M, Vesikansa A et al. Endogenous activation of kainate receptors regulates glutamate release and network activity in the developing hippocampus. J Neurosci 2005; 25:4473-4484.
38. Kidd FL, Coumis U, Collingridge GL et al. A presynaptic kainate receptor is involved in regulating the dynamic properties of thalamocortical synapses during development. Neuron 2002; 34:635-646.
39. Li H, Chen A, Xing G et al. Kainate receptor-mediated heterosynaptic facilitation in the amygdala. Nat Neurosci 2001; 4:612-620.
40. Crowder TL, Weiner JL. Functional characterization of kainate receptors in the rat nucleus accumbens core region. J Neurophysiol 2002; 88:41-48.
41. Casassus G, Mulle C. Functional characterization of kainate receptors in the mouse nucleus accumbens. Neuropharmacology 2002; 42:603-611.
42. Kerchner GA, Wilding TJ, Li P et al. Presynaptic kainate receptors regulate spinal sensory transmission. J Neurosci 2001; 21:59-66.
43. Delaney AJ, Jahr CE. Kainate receptors differentially regulate release at two parallel fiber synapses. Neuron 2002; 36:475-482.
44. Rodríguez-Moreno, Sihra TS. Kainate receptors with a metabotropic modus operandi. Trends Neurosci 2007; 30:630-637.
45. Rodríguez-Moreno A, Sihra TS. Metabotropic actions of kainate receptors in the CNS. J Neurochem 2007; 103:2121-2135.

Chapter 5

Metabotropic Actions of Kainate Receptors in the Regulation of I_{sAHP} and Excitability in CA1 Pyramidal Cells

Zara Melyan* and Howard V. Wheal

Abstract

Kainate receptors (KARs) mediate postsynaptic responses in CA3 pyramidal cells and CA1 interneurones in the hippocampus. In CA1 pyramidal cells knockout studies have inidcated the presence of functional GluR6-containing KARs, however in this region they made no ionotropic contribution to the synaptic responses. In the meantime, a metabotropic function was reported for presynaptic KARs modulating transmitter release in CA1. We examined the possibility that KARs in CA1 pyramidal cells have a metabotropic function. Kainate is known to inhibit a slow afterhyperpolarization current that regulates excitability in hippocampus and can be modulated by a number of G protein coupled receptors. We showed that KARs activation reduces slow afterhyperpolarization current in CA1 pyramidal cells via metabotropic action and elucidated the transduction mechanism(s) underlying this action.

Introduction

Significant progress in our understanding of kainate receptor functions has followed from the development of the selective AMPA receptor antagonists (GYKI52466 and GYKI53655) that enabled discrimination between AMPA and kainate subtypes of glutamate receptors.

The glutamatergic excitatory postsynaptic currents (EPSCs) were first shown to contain a kainate receptor component in pyramidal cells of the hippocampal CA3 region.[1-2] Synaptic activation of KARs has now been described for several synapses throughout the brain and spinal cord, including CA1 hippocampal interneurons.[3-4] In CA1 pyramidal neurons knockout studies have demonstrated the presence of functional GluR6-containing kainate receptors that produce inward currents,[5] however they made no detectable ionotropic contribution to synaptic responses in these cells. In the meantime, KARs classically described as ionotropic were reported to have a metabotropic function in CA1 interneurons.[6] The latter finding inspired us to investigate the possibility that the kainate receptors that failed to contribute to synaptic responses in CA1, instead, have a metabotropic function.

One of the most prominent postsynaptic targets of metabotropic transmitter action in the hippocampus is the slow afterhyperpolarization (sAHP) which regulates action potential firing frequency and is responsible for the spike-frequency adaptation. It has a slow rising phase, lasts several seconds and is generated by a voltage-independent, Ca^{2+}-dependent K^{+} current.[7] In CA1 pyramidal neurones the sAHP current (I_{sAHP}) is activated proportionally to the number and

*Corresponding Author: Zara Melyan—Circadian and Visual Neuroscience Group, Nuffield Laboratory of Ophthalmology, University of Oxford, Levels 5 and 6 West Wing, The John Radcliffe Hospital, Headley Way, Oxford OX3 9DU, UK. Email: zara.melyan@eye.ox.ac.uk

Kainate Receptors: Novel Signaling Insights, edited by Antonio Rodríguez-Moreno and Talvinder S. Sihra.

frequency of action potentials within a burst and provides a negative feedback mechanism to repetitive spiking and hyperexcitability.[8-9] A number of neurotransmitters like acetylcholine, noradrenaline or glutamate inhibit sAHP via G-protein-coupled receptors.[10] The transduction mechanism for acetylcholine induced inhibition involves the Ca^{2+}-calmodulin dependent protein kinase II,[11] whereas noradrenaline action has been shown to be mediated by cAMP and protein kinase A.[12-13]

The first observation of the effect of kainic acid on the sAHP was reported in 1986.[14] Intracerebroventricular or intraperitoneal kainate infusions resulted in long-lasting seizures and hyperexcitability making kainate a perfect model for studying temporal lobe epileptiform activity. This action of kainate is accompanied by neuronal loss and glial proliferation in the hippocampal tissue. It has been shown that in the kainic acid-lesioned hippocampus, the sAHP is reduced compared to control. The resulting increase in excitability may contribute to the seizure-like activity that is caused by kainate, although the cellular mechanism of this effect of kainate is not fully understood. Here we describe some key experiments that show that acute application of kainic acid as well as endogenous glutamate release reduce I_{sAHP} in CA1 pyramidal cells via metabotropic actions of kainate receptors and provide a basis for understanding of transduction mechanisms behind this action.

Direct Postsynaptic Activation of KAR Inhibits I_{sAHP} on CA1 Pyramidal Cell

Experiments were performed on 14- to 19-day-old rat hippocampal slices. Whole-cell patch-clamp recordings were made from stratum pyramidale using $KMeSO_4$-containing pipettes. I_{sAHP} was recorded in CA1 pyramidal neurons voltage clamped at holding potentials from −50 to −65 mV following 80 ms voltage steps to −10 mV applied every 20 s. The effect of synaptically released glutamate was assessed by comparing sAHP currents induced before and after a train of stimuli delivered via a stimulating electrode.[15-16]

Bath application of 200 nM kainate caused a long-lasting inhibition of I_{sAHP} of 34% ± 6% (n = 7; Fig. 1A).[15] The dose-response curve shows that the effect of kainate was concentration-dependent reaching a plateau of 34% at the concentration of 100 nM with an IC_{50} = 15 nM (Fig. 1B). In order to test whether the effect of kainate was exclusively mediated by KARs, we used a cocktail of antagonists: for the NMDA receptor (100 μM DL-AP5), AMPA receptor (100 μM GYKI52466), metabotropic glutamate recepotors (1 mM MCPG and 250 μM MSOP), $GABA_A$ receptor (100 μM picrotoxin), $GABA_B$ receptor (200 μM 2-OH-saclofen), muscarinic actetylcholine receptor (1 μM atropine sulfate), opioid receptor (10 μM naloxone), cannabinoid CB_1 receptor (2 μM AM 251) and adenosine receptor (0.1 μM DPCPX). In the presence of these receptor blockers the effect of kainate was indistinguishable from the control values (n = 5; Fig. 1C). To test if I_{sAHP} inhibition was an indirect effect, caused by action potential-dependent transmitter release, we applied kainate after blocking the action potentials with 1 μM tetrodotoxin (TTX). Under these conditions the activation of kainate receptors was still able to produce an inhibition of I_{sAHP} of 35% ± 3% (n = 8; Fig. 1D).

Kainate-induced I_{sAHP} inhibition was completely abolished by prior application of 20 μM CNQX indicating a requirement for AMPA/kainate receptors (n = 6; Fig. 2A). In contrast, application of 20 μM CNQX after kainate (n = 3; Fig. 2B) did not block I_{sAHP} inhibition, showing that the long-lasting effect is unlikely to be due to persistent receptor activation or slow washout of the kainate but may, instead, reflect the action of a second messenger. A summary histogram for different agonists and antagonists of AMPA and kainate receptors is shown on Figure 2C. Both agonists of KAR, kainate and domoate (200 nM),[17] caused I_{sAHP} inhibition, while the potent AMPA receptor agonist (S)-5-fluorowillardiine (300 nM)[18-19] showed no effect. The nonselective AMPA/kainate receptor antagonist CNQX (20 μM) blocked the effect of kainate, yet the selective AMPA receptor blocker GYKI52466 (100 μM)[20] was ineffective at 5-10 times the reported IC_{50} values.[21] At concentrations known to be effective against GluR5 subunits,[22] ATPA (2 μM) did not produce any changes in I_{sAHP} amplitude. These data are consistent with a direct action of kainate

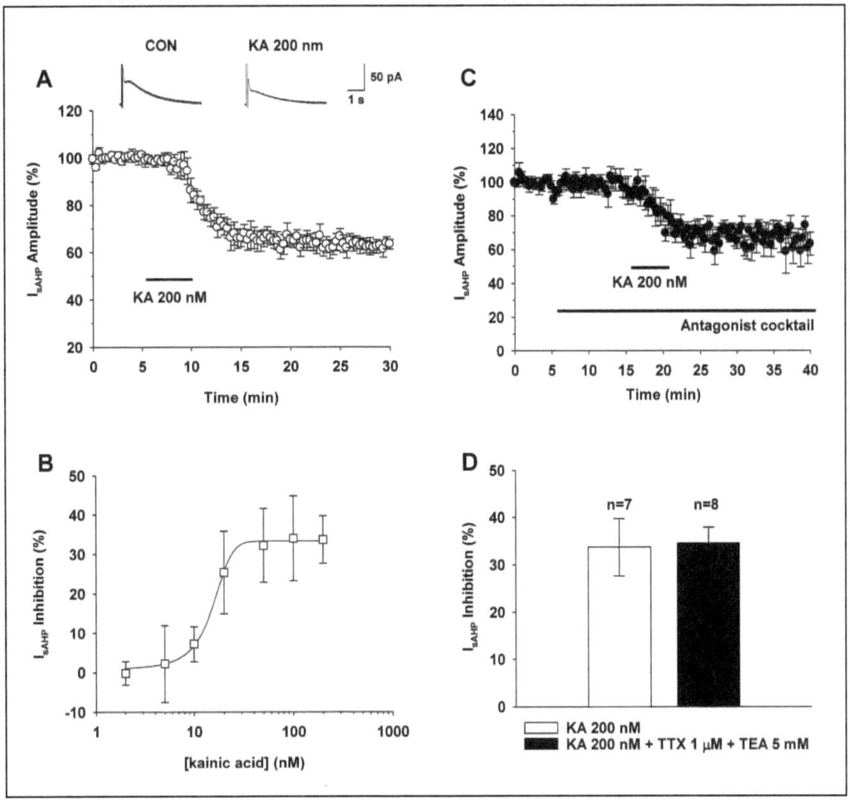

Figure 1. Kainate inhibits I_{sAHP} by a direct action on CA1 pyramidal cells. A) Summary time course of I_{sAHP} inhibition caused by 200 nM kainate application from 7 experiments (mean ± SEM). The traces illustrate data from one neuron (averages of 15 trials) before and after 200 nM kainate application. B) Dose-response curve for kainate-induced inhibition of I_{sAHP}. Each symbol represents average of at least 4 cells. The data were fitted with a logistic curve which gives an IC_{50} = 15 nM. C) Summary time course of I_{sAHP} inhibition from 5 experiments (mean ± SEM) recorded in the presence of the cocktail of antagonists. D) Comparison of the 200 nM kainate evoked inhibition with and without 1 μM TTX + 5 mM TEA. The data presented are the average percentage inhibition of I_{sAHP} (mean ± SEM). Reproduced from: Melyan Z et al. Neuron 2002; 34:107-114;[15] ©2002 with permission from Elsevier.

on CA1 pyramidal cells, acting via GluR6-containing kainate receptors to modulate I_{sAHP}. This was confirmed recently by knockout studies which have shown that GluR6−/− or KA2−/−, but not GluR5−/− mice, lack kainate-induced inhibition of the slow AHP.[23-24]

We investigated whether synaptic activation of KARs can produce inhibition of I_{sAHP} similar to that evoked by exogenously applied kainate.[16] In order to isolate KAR activity all the experiments were carried out in the presence of the cocktail of antagonists (see above), which abolished the mixture of EPSCs and IPSCs that could be evoked in CA1 pyramidal cell by single-pulse stimulation of Schaffer/commissural afferents in control conditions. Trains of synaptic stimuli (5-pulse 100-Hz) previously shown to activate KAR[25] produced 37 ± 1% inhibition of I_{sAHP} amplitude (n = 8; Fig. 3). The I_{sAHP} inhibition was not reversible and had a time course similar to that recorded after 200 nM kainate application. The glutamate uptake

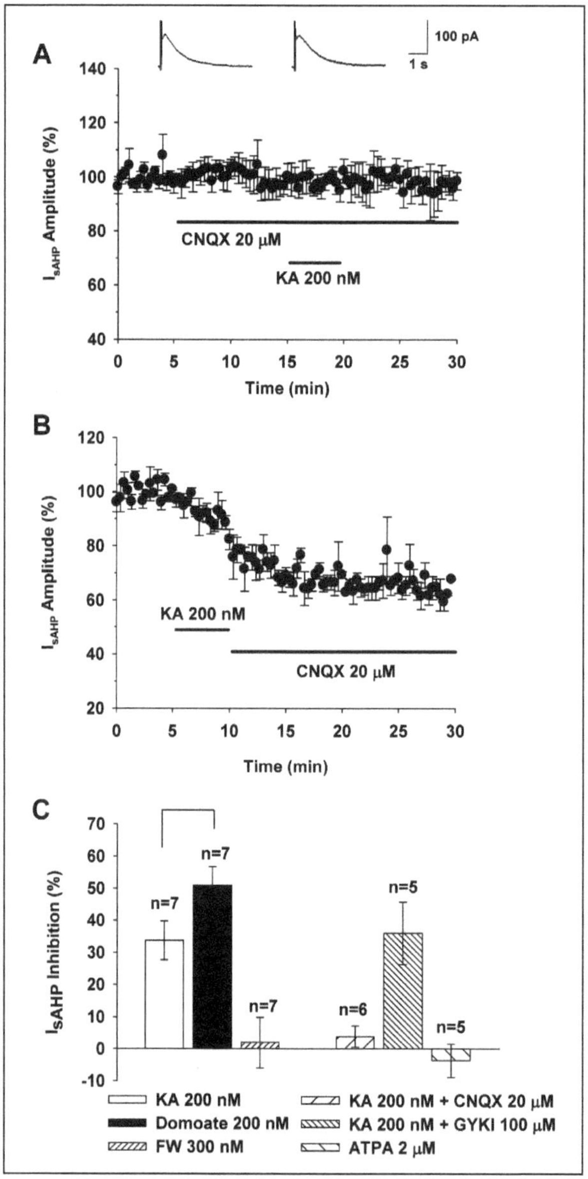

Figure 2. I_{sAHP} inhibition is mediated by GLuR6- containing KAR activation. A) Prior application of 20 µM CNQX blocked the effect of kainate. Summary time course of I_{sAHP} inhibition (n = 6). The traces (averages of 15 trials) were recorded in the presence of 20 µM CNQX before and after 200 nM kainate was added to the solution. B) 20 µM CNQX applied after 200 nM kainate did not relieve I_{sAHP} inhibition (n = 6). C) Summary histogram of the effects of AMPA and kainate receptor agonists and antagonists on I_{sAHP}: FW, (S)-5-fluorowillardiine (AMPA agonist); GYKI (AMPA antagonist); ATPA (GluR5 containing KAR agonist). The bars are the average percentage inhibition of I_{sAHP} (mean ± SEM). Reproduced from: Melyan Z et al. Neuron 2002; 34:107-114;[15] ©2002 with permission from Elsevier.

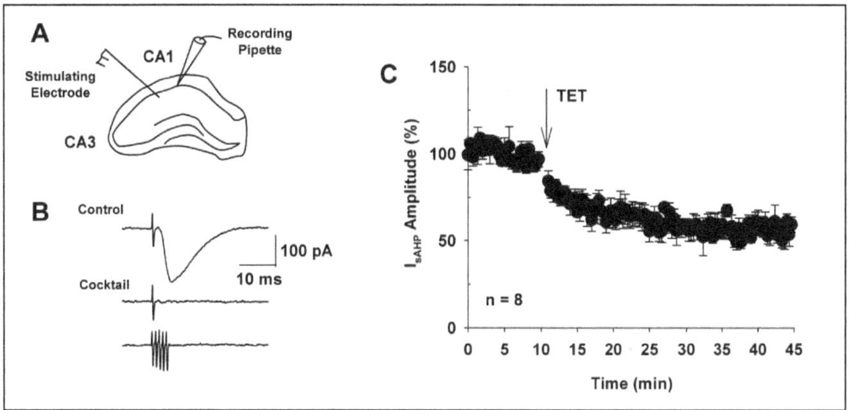

Figure 3. Synaptically released glutamate mimics kainate induced I_{sAHP} inhibition in CA1 pyramidal cells. A) Schematic illustration of the hippocampal slice showing the positioning of the electrodes. B) Single-pulse stimulation of Schaffer/commissural afferents evoked an EPSC in a CA1 pyramidal cell that was blocked by the antagonist mixture. A 5 pulse, 100 Hz stimulation did not produce any synaptic response in the presence of the cocktail. C) A 5 pulse, 100 Hz stimulation reduced the amplitude of I_{sAHP}. Summary time course of I_{sAHP} inhibition from 8 experiments (mean ± SEM). Reproduced from: Melyan Z et al. J Neurosci 2004; 24(19):4530-4534;[16] ©2004 with permission from the Society for Neuroscience.

blocker TBOA (50 μM) potentiated the effect of synaptic stimulation on I_{sAHP} (n = 8; Fig. 4A), consistent with a prolongation of the transient synaptically-released glutamate. Increasing the number of stimuli in the pulse to 10 or 20 also increased I_{sAHP} inhibition (Fig. 4B). The I_{sAHP} inhibition recorded in the presence of TBOA and the cocktail of blockers was effectively blocked by prior application of 20 μM CNQX (n = 7; Fig. 4C), confirming that it was mediated by kainate receptor activation. Previous activation of KARs by bath-applied kainate occluded the action of the released glutamate showing that the synaptic stimulation and exogenous kainate act on the same population of KARs (n = 6; Fig. 4D).

KARs Responsible for I_{sAHP} Inhibition Involve Metabotropic Action

The modulation of I_{sAHP} by either application of kainate or synaptically released glutamate was not accompanied by any changes in membrane conductance. This implies that the ionotropic properties of KAR are unlikely to account for these effects of I_{sAHP}. To test for metabotropic functions we used an inhibitor of pertussis toxin-sensitive G proteins, N-ethylmaleimide (NEM)[26-27] and protein kinase C inhibitor (PKC), calphostin C.

Bath applied NEM (50 μM) abolished kainate-induced I_{sAHP} inhibition (n = 7). PKC has long been known to inhibit the sAHP,[28] although the transmitter linked to this action was not identified. In 1998 Rodriguez-Moreno and Lerma reported that the metabotropic action of presynaptic KARs involved a PKC activation.[6] This prompted us to perform similar tests which demonstrated that preincubation of the slices with 1 μM calphostin C for 2-4 hours blocked the action of kainate to inhibit I_{sAHP} (n = 10), while subsequent application of 10 μM noradrenaline, known to reduce I_{sAHP} via PKA,[29] still blocked I_{sAHP} ruling out a broad spectrum kinase inhibition and indicating, instead, a specific requirement for PKC in the action of kainate (Fig. 5A). Similarly, preincubation of the slices with 1 μM calphostin C for 2-4 hours prior recording prevented I_{sAHP} inhibition by glutamate released after 5 pulse 100 Hz stimulation of excitatory afferents (n = 8; Fig. 5B).

Taken together, these data show that exogenous kainate application as well as synaptically released glutamate inhibit I_{sAHP} acting through metabotropic kainate receptors on CA1 pyramidal cells.

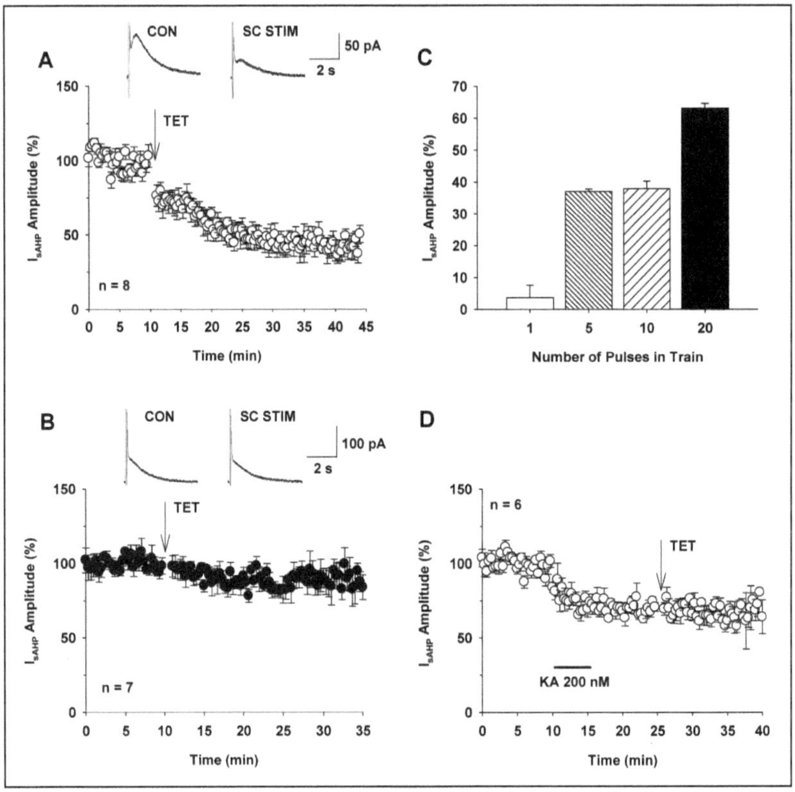

Figure 4. I_{sAHP} inhibition is mediated by glutamate acting on kainate receptors. A) Block of glutamate uptake increased the effect of tetanic stimulation. Sample traces were obtained before and after a tetanus in the same neuron in the presence of the glutamate uptake inhibitor TBOA (50 μM; averages of 15 trials). Summary time course of I_{sAHP} inhibition recorded in the presence of TBOA from 8 experiments (mean ± SEM) is shown in the graph. B) Previous application of the AMPA/kainate receptor antagonist CNQX completely abolished the effect of synaptically released glutamate in the presence of TBOA. Sample traces were obtained in the presence of 20 μM CNQX and TBOA before and after a tetanic stimulation in the same neuron (averages of 15 trials). Summary time course of I_{sAHP} amplitude recorded in the presence of 20 μM CNQX from seven experiments (mean ± SEM) is plotted in the graph. C) Summary histogram showing the relation between the number of synaptic stimuli and the inhibition of I_{sAHP}. D) Summary time course of I_{sAHP} inhibition recorded during 200 nm kainate application, followed by 5 pulse, 100 Hz stimulation (n = 6). Kainate application occluded the inhibitory action of synaptic glutamate release. Reproduced from: Melyan Z et al. J Neurosci 2004; 24:4530-4534;[16] ©2004 with permission from the Society for Neuroscience.

Signal Transduction Mechanisms for Metabotropic KAR-Induced Inhibition of I_{sAHP} in CA1 Pyramidal Cells

The involvement of G proteins in the KAR-induced inhibition of I_{sAHP} was originally demonstrated using bath applied NEM. The follow-up experiments[30] using intracellular exposure to pertussis toxin (2.5 μg ml−1, >10 min)[31] or the $G\alpha_{i/o}$ blocker, NF023 (10 μM),[32] confirmed that the metabotropic KARs acted through $G\alpha_{i/o}$ proteins, as these compounds preferentially depressed kainate action leaving β-adrenergic inhibition of the sAHP unaffected.

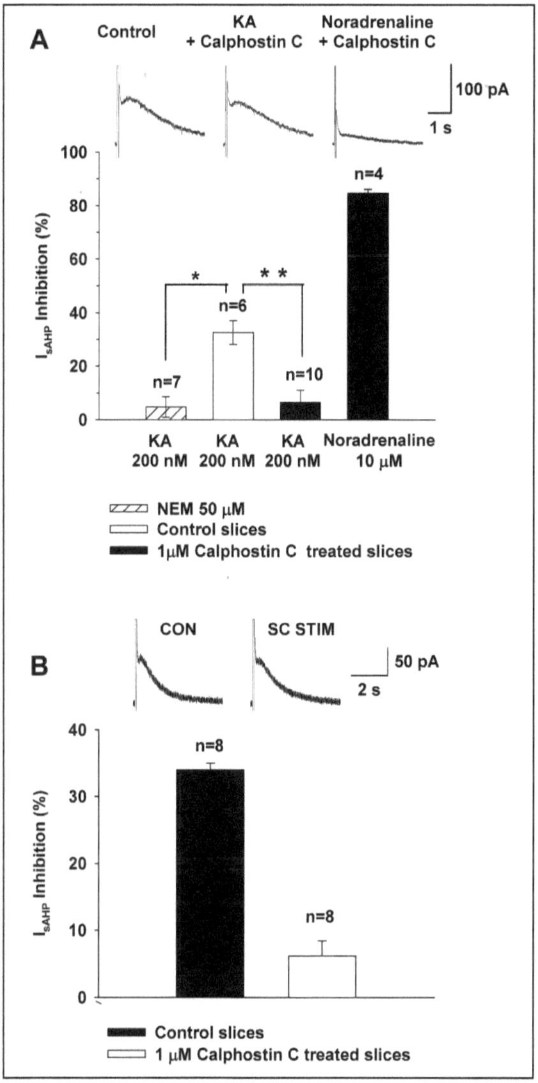

Figure 5. KAR-induced inhibition of I_{sAHP} requires PKC. A) Sample traces (averages of 15 trials) were recorded in calphostin C-treated slices before and after kainate application and after the following application of noradrenaline. The histogram shows the average percentage inhibition of I_{sAHP} (mean ± SEM). 200 nM kainate inhibited I_{sAHP} in control slices (white bar). In calphostin C (1 μM) treated slices (black bars) kainate-induced inhibition was significantly reduced ($p < 0.002$, unpaired t-test), subsequent application of 10 μM noradrenaline blocked I_{sAHP}. Part A is reproduced from: Melyan Z et al. Neuron 2002; 34:107-114;[15] © 2002 with permission from Elsevier. B) Sample traces (averages of 15 trials) were recorded in calphostin C-treated slices before and after tetanic stimulation. The histogram shows the average percentage inhibition (mean ± SEM) of I_{sAHP}. A train of five stimuli inhibited I_{sAHP} in control slices (white bar). In calphostin C-treated slices (1 μM; black bar) kainate-induced inhibition was reduced significantly. Part B reproduced from: Melyan Z et al. J Neurosci 2004; 24(19):4530-4534;[16] ©2004 with permission from the Society for Neuroscience.

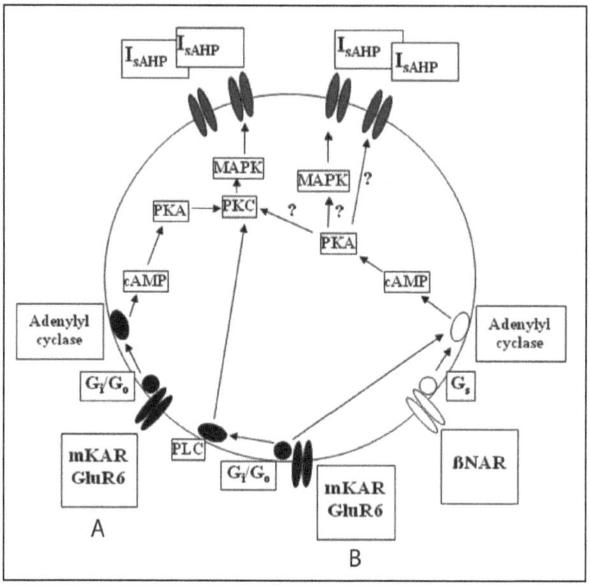

Figure 6. A model for the action of postsynaptic metabotropic kainate receptors on I_{sAHP} in CA1 pyramidal cell. A) shows a hypothetic transduction cascade where PKC acts downstream of PKA. B) shows another hypothetic pathway where PKC acts independently of PKA. βNAR, β-adrenergic receptor; mKAR GluR6, GluR6 containing metabotropic KAR; MAPK, mitogen-activated protein kinase; PLC, phospholipase C; PKA, cAMP-depenent protein kinase; PKC, protein kinase.

Noradrenaline is known to block sAHP via protein kinase A and cAMP. The fact, that the PKC inhibitor calphostin C, as well as the pertussis toxin and NF023, prevented I_{sAHP} inhibition in CA1 pyramidal cells without affecting the subsequent I_{sAHP} inhibition caused by activation of noradrenergic β receptors, suggested an independence of the two transduction pathways. However, surprisingly two distinct PKA inhibitors, Rp-cAMPs (50 μM) and H89 (10 μM) applied intracellularly,[33-34] have been found to prevent the action of kainate on I_{sAHP}.[30] The adenylyl cyclase inhibitor DDA (10 μM)[35] attenuated both KAR and β-adrenergic receptor I_{sAHP} inhibition in CA1 showing that there was also a corresponding requirement for cAMP production. Taken together, these data indicate that activation of adenylyl cyclase/PKA is a common requirement for metabotropic actions of KARs and β-adrenergic receptors; however, the involvement of $G\alpha_{i/o}$ and PKC is specific to KARs. The subsequent experiments[30] showed that the action of PKC on the sAHP was unaffected by inhibition of PKA, implying that PKC can act either downstream of (Fig. 6, pathway A) or independently from PKA (Fig. 6, pathway B) in this transduction cascade.

In CA1 cells, both PKC and β-adrenergic receptors, acting via PKA, have been shown to couple to MAP kinase.[36] MAP kinases were known to be involved in late stage of long-term potentiation,[37-39] therefore, MAP kinase cascade represented a possible effector mechanism for the KAR induced long-lasting inhibition of I_{sAHP}. Recent results[30] showed that the ability of bath applied kainate to inhibit the sAHP was blocked by either of two structurally distinct MEK inhibitors U0126 (10 μM) and PD098059 (10 μM)[40] included in the pipette. Interestingly, when a MEK inhibitor U0126 (10 μm) was bath applied after a stable control period, the amplitude of I_{sAHP} gradually increased, while the inclusion of activated MAP kinase within the recording pipette caused a gradual loss of the current, indicating a bidirectional mechanism of regulation of neuronal excitability by metabotropic KARs through MAP kinase activation.

Conclusion

- Activation of KARs by either low concentrations of kainate or synaptically released glutamate inhibit I_{sAHP} in CA1 pyramidal cells
- This is a direct postsynaptic effect mediated by a high affinity GluR6-containing kainate receptor
- This is a metabotropic type effect linked to a pertussis toxin sensitive G protein and PKC signalling cascade
- KAR induced I_{sAHP} inhibition in CA1 shows a requirement for PKA and cAMP
- PKC acts either downstream of, or independently from, PKA
- MAP kinase cascade is an effector mechanism for the KAR induced long-lasting inhibition of I_{sAHP}

Acknowledgement

We would like to thank Dr. Barrie Lancaster for his invaluable input to this work and his constructive review of the manuscript.

References

1. Vignes M, Collingridge GL. The synaptic activation of kainate receptors. Nature 1997; 388:179-182.
2. Castillo PE, Malenka RC, Nicoll RA. Kainate receptors mediate a slow postsynaptic current in hippocampal CA3 neurons. Nature 1997; 388:182-186.
3. Cossart R, Esclapez M, Hirsch JC et al. GluR5 kainate receptor activation in interneurons increases tonic inhibition of pyramidal cells. Nat Neurosci 1998; 1:470-478.
4. Frerking M, Malenka RC, Nicoll RA. Synaptic activation of kainate receptors on hippocampal interneurons. Nat Neurosci 1998; 1:479-486.
5. Bureau I, Bischoff S, Heinemann SF et al. Kainate receptor-mediated responses in the CA1 field of wild-type and GluR6-deficient mice. J Neurosci 1999; 19:653-663.
6. Rodríguez-Moreno A, Lerma J. Kainate receptor modulation of GABA release involves a metabotropic function. Neuron 1998; 20:1211-1218.
7. Lancaster B, Adams PR. Calcium-dependent current generating the afterhyperpolarization of hippocampal neurons. J Neurophysiol 1986; 55:1268-1282.
8. Madison DV, Nicoll RA. Control of the repetitive discharge of rat CA1 pyramidal neurons in vitro. J Physiol 1984; 354:319-331.
9. Traub RD, Miles R, Jefferys JGR. Synaptic and intrinsic conductances shape picrotoxin-induced synchronized after-discharges in the guinea-pig hippocampal slice. J Physiol 1993; 461:525-547.
10. Nicoll RA. The coupling of neurotransmitter receptors to ion channels in the brain. Science 1988; 241:545-551.
11. Müller W, Petrozzino JJ, Griffith LC et al. Specific involvement of Ca^{2+}-calmodulin kinase II in cholinergic modulation of neuronal responsiveness. J Neurophysiol 1992; 68:2264-2269.
12. Madison DV, Nicoll RA. Actions of noradrenaline recorded intracellularly in rat hippocampal CA1 pyramidal neurons, in vitro. J Physiol 1986; 372:221-244.
13. Pedarzani P, Storm JF. PKA mediates the effects of monoamine transmitters on the K^+ current underlying the slow spike frequency adaptation in hippocampal neurons. Neuron 1993; 11:1023-1035.
14. Ashwood TJ, Lancaster B, Wheal HV. Intracellular electrophysiology of CA1 pyramidal neurons in slices of the kainic acid lesioned hippocampus of the rat. Exp Brain Res 1986; 62:189-198.
15. Melyan Z, Wheal HV, Lancaster B. Metabotropic-mediated kainite receptor regulation of I_{sAHP} and excitability in pyramidal cells. Neuron 2002; 34:107-114.
16. Melyan Z, Lancaster B, Wheal HV. Metabotropic regulation of intrinsic excitability by synaptic activation of kainate receptors. J Neurosci 2004; 24:4530-4534.
17. Debonnel G, Beauchesne L, de Montigny C. Domoic acid, the alleged "mussel toxin," might produce its neurotoxic effect through kainate receptor activation: an electrophysiological study in the dorsal hippocampus. Can J Physiol Pharmacol 1989; 67:29-33.
18. Wong LA, Mayer ML, Jane DE et al. Willardiines differentiate agonist binding sites for kainate-versus AMPA-preferring glutamate receptors in DRG and hippocampal neurons. J Neurosci 1994; 14:3881-3897.
19. Jane DE, Hoo K, Kamboj R et al. Synthesis of willardiine and 6-azawillardiine analogs: pharmacological characterization on cloned homomeric human AMPA and kainate receptor subtypes. J Med Chem 1997; 40:3645-3650.
20. Paternain AV, Morales M, Lerma J. Selective antagonism of AMPA receptors unmasks kainate receptor-mediated responses in hippocampal neurons. Neuron 1995; 14:185-189.

21. Bleakman D, Ballyk BA, Schoepp DD et al. Activity of 2,3-benzodiazepines at native rat and recombinant human glutamate receptors in vitro: stereospecificity and selectivity profiles. Neuropharmacology 1996; 35:1689-1702.
22. Clarke VR, Ballyk BA, Hoo KH et al. A hippocampal GluR5 kainate receptor regulating inhibitory synaptic transmission. Nature 1997; 389:599-603.
23. Fisahn A, Contractor A, Traub RD et al. Distinct roles for the kainate receptor subunits GluR5 and GluR6 in kainate-induced hippocampal gamma oscillations. J Neurosci 2004; 24:9658-9668.
24. Ruiz A, Sachidhanandam S, Utvik JK et al. Distinct subunits in heteromeric kainate receptors mediate ionotropic and metabotropic function at hippocampal mossy fiber synapses. J Neurosci 2005; 25:11710-11718.
25. Min MY, Melyan Z, Kullmann DM. Synaptically released glutamate reduces gamma-aminobutyric acid (GABA)ergic inhibition in the hippocampus via kainate receptors. Proc Natl Acad Sci USA 1999; 96:9932-9937.
26. Heuss C, Scanziani M, Gahwiler BH et al. G-protein-independent signaling mediated by metabotropic glutamate receptors. Nat Neurosci 1999; 2:1070-1077.
27. Frerking M, Schmitz D, Zhou Q et al. Kainate receptors depress excitatory synaptic transmission at CA3-CA1 synapses in the hippocampus via a direct presynaptic action. J Neurosci 2001; 21:2958-2966.
28. Malenka RC, Madison DV, Andrade R et al. Phorbol esters mimic some cholinergic actions in hippocampal pyramidal neurons. J Neurosci 1986; 6:475-480.
29. Pedarzani P, Storm JF. PKA mediates the effects of monoamine transmitters on the K^+ current underlying the slow spike frequency adaptation in hippocampal neurons. Neuron 1993; 11:1023-1035.
30. Grabauskas G, Lancaster B, O'Connor V et al. Protein kinase signalling requirements for metabotropic action of kainate receptors in rat CA1 pyramidal neurones. J Physiol 2007; 579(2):363-373.
31. Katada T, Ui M. Direct modification of the membrane adenylate cyclase system by islet-activating protein due to ADP-ribosylation of a membrane protein. Proc Natl Acad Sci USA 1982; 79:3129-3133.
32. Freissmuth M, Boehm S, Beindl W et al. Suramin analogues as subtype-selective G protein inhibitors. Mol Pharmacol 1996; 49:602-611.
33. Chijiwa T, Mishima A, Hagiwara M et al. Inhibition of forskolin-induced neurite outgrowth and protein phosphorylation by a newly synthesized selective inhibitor of cyclic AMP-dependent protein kinase, N-[2-(p-bromocinnamylamino) ethyl]-5-isoquinolinesulfonamide (H-89), of PC12D pheochromocytoma cells. J Biol Chem 1990; 265:5267-5272.
34. Dostmann WR, Taylor SS, Genieser HG et al. Probing the cyclic nucleotide binding sites of cAMP-dependent protein kinases I and II with analogs of adenosine 3′,5′-cyclic phosphorothioates. J Biol Chem 1990; 251:10484-10491.
35. Holgate ST, Lewis RA, Austen KF. Role of adenylate cyclase in immunologic release of mediators from rat mast cells: agonist and antagonist effects of purine- and ribose-modified adenosine analogs. Proc Natl Acad Sci USA 1980; 77:6800-6804.
36. Roberson ED, English JD, Adams JP et al. The mitogen-activated protein kinase cascade couples PKA and PKC to cAMP response element binding protein phosphorylation in area CA1 of hippocampus. J Neurosci 1999; 19:4337-4348.
37. English JD, Sweatt JD. Activation of p42 mitogen-activated protein kinase in hippocampal long term potentiation. J Biol Chem 1996; 271:24329-24332.
38. Coogan AN, O'Leary DM, O'Connor JJ. P42/44 MAP kinase inhibitor PD98059 attenuates multiple forms of synaptic plasticity in rat dentate gyrus in vitro. J Neurophysiol 1999; 81:103-110.
39. Winder DG, Martin KC, Muzzio IA et al. ERK plays a regulatory role in induction of LTP by theta frequency stimulation and its modulation by beta-adrenergic receptors. Neuron 1999; 24:715-726.
40. Lubin FD, Johnston LD, Sweatt JD et al. Kainate mediates nuclear factor-kappa B activation in hippocampus via phosphatidylinositol-3 kinase and extracellular signal-related protein kinase. Neuroscience 2005; 133:969-981.

Chapter 6

Kainate Receptors with a Metabotropic Signature Enhance Hippocampal Excitability by Regulating the Slow After-Hyperpolarization in CA3 Pyramidal Neurons

Arnaud Ruiz*

Abstract

Most of our knowledge of the synaptic function of kainate receptors stems from a detailed analysis of synaptic transmission between dentate granule cells and CA3 pyramidal neurons, where kainate receptors mediate a slow excitatory current with integrative properties ideally suited for repetitive neuronal firing. Besides this well characterized ionotropic effect of kainate receptors, they can also enhance neuronal excitability by inhibiting the slow Ca^{2+} activated K^+ current I_{sAHP} via a G-protein coupled mechanism. This phenomenon is associated with Ca^{2+} mobilization and protein-kinase activation and ultimately leads to modulation of ion channels responsible for intrinsic electrical properties such as firing adaptation. The significance for CNS function of these newly emerging metabotropic kainate receptors is poorly understood and as yet proteomic analysis of kainate receptors has yielded little information on signaling molecules associated with the kainate receptor ionophore. This chapter covers the key findings that have led to the proposal that high-affinity postsynaptic kainate receptors trigger a form of metabotropic signaling regulating I_{sAHP} and neuronal firing in CA3 hippocampal neurons.

Introduction

Kainate receptors (KARs) are ionotropic glutamate receptors composed of a distinct family of subunits with strong sequence homology to α-amino-3-hydroxy-5-methyl-4-isoxazole propionate receptors (AMPARs). Although recent advances in molecular cloning have shown that AMPARs and KARs form separate entities, the function of KARs at central excitatory synapses has remained elusive. Evidence for separate KARs and AMPARs came from the differential sensitivity of spinal cord C-fibres to kainate and quisqualate.[1] Pharmacological manipulation of kainate receptors in the presence of 2, 3-benzodiazepines acting as AMPA receptor antagonists,[2-4] combined with the use of genetically-engineered mice strains where specific kainate receptor genes have been inactivated, has eased the tedious process of identifying the function of individual KAR subunits. In general, KARs generate excitatory postsynaptic currents (EPSCs) albeit with much slower kinetics

*Arnaud Ruiz—Department of Pharmacology, The School of Pharmacy, University of London, Brunswick Square, London WC1N 1AX, UK. Email: arnaud.ruiz@pharmacy.ac.uk

Kainate Receptors: Novel Signaling Insights, edited by Antonio Rodríguez-Moreno and Talvinder S. Sihra.

and smaller amplitudes than those mediated by AMPARs.[5-11] The slow KAR-mediated EPSC at hippocampal mossy fiber synapses provides unique additive properties of synaptic inputs arising from dentate granule cells projecting to CA3 pyramidal neurons.[12-14] However, a more widespread finding in the hippocampus has been that KAR activation by exogenous agonist application or by endogenous release of glutamate can lead to neuromodulation of neuronal function, both through complex effects on neurotransmitter release,[15-23] or by means of persistent changes of neuronal excitability. In keeping with this view, several reports have now demonstrated that KARs can affect neuronal excitability by modulating the size of the slow after-hyperpolarization generated by a voltage-independent, Ca^{2+} dependent K^+ current (I_{sAHP}). This inhibition of I_{sAHP} was prevented by inhibitors of G-protein dependent cascades and was mimicked by a high frequency train of electrical stimuli designed to activate synaptically CA1 or CA3 pyramidal neurons.[24-28] Thus, overwhelming evidence has accumulated whereby KARs can exert two distinct functions, namely a well characterized ionotropic action through a cationic conductance,[29] or an indirect regulation of ion channels mediated by G-protein activation and intracellular signaling pathways.[30]

This chapter reviews the evidence that synaptically-released glutamate can enhance pyramidal neuron excitability via a pathway involving KARs and G-protein activation at hippocampal excitatory synapses. An important question derived from this observation is to ask which KAR subunit(s) mediate these effects and what are the interacting partners linking KARs to other ion channels that have known actions on neuronal excitability. We will begin by examining current ideas about KAR subunit expression in the hippocampus as well as data derived from functional expression studies because they provide useful information on receptor distribution and trafficking or subunit assembly in native receptors, which, altogether, are critical determinants of the processes underlying changes of cellular excitability. Subsequent sections will cover the pharmacological paradigms aiming at describing the effect of KAR activation on I_{sAHP} in pyramidal neurons in different KAR knockout mice, with the common idea that synaptically-released glutamate can mimic the effect of exogenous agonists on I_{sAHP}. Finally, different signaling pathways linking KARs and the putative ion channel(s) underlying I_{sAHP} will be brought together and speculation will be made on the *modus operandum*[30] by which KARs could account both for ionotropic and metabotropic actions in CA3 pyramidal neurons.

High-Affinity Postsynaptic KAR Subunits in the Hippocampus

At present, the role of high-affinity KAR subunits on synaptic transmission remains controversial. The main KAR subunits GluK1, GluK2 and GluK3 (previously termed GluR5, GluR6 and GluR7) form functional KARs when expressed in recombinant expression systems.[13,14,31-37] GluK3 is thought to localize presynaptically at mossy fiber synapses where it mediates an autoreceptor function.[38] GluK4 and GluK5 subunits, formerly known as KA1 and KA2, do not form functional homomeric receptor channels but rather co-assemble with GluK1-3 to modulate pharmacological and biophysical properties of recombinant KARs.[13,39,40] Co-assembly of GluK5 with GluK2 does not affect the efficacy by which KARs bind glutamate.[32] However, currents mediated by recombinant GluK2/GluK5 receptors, but not GluK2 receptors, decay with a time course similar to KAR-EPSCs in response to brief glutamate pulses. These gating features could contribute to the slow KAR-EPSC decay kinetics observed at mossy fiber—CA3 pyramidal neuron synapses.[12-14] The GluK5 subunit is expressed ubiquitously in nearly all native receptor populations, suggesting that it is a constituent of most heteromeric neuronal KAR complexes.[41] Moreover, GluK5 is not sufficient for the functional expression of postsynaptic KARs in CA3 pyramidal cells since KAR-EPSCs are absent in $GluK2^{-/-}$ mice.[28,42,43] Indeed, no change in KAR-EPSCs amplitude has been reported in $GluK5^{-/-}$ mice,[12,44] suggesting that GluK5 is not required for ionotropic signaling mediated by postsynaptic KARs at hippocampal mossy fiber synapses. The recent introduction of a double knock-out mice $GluK4^{-/-}$/$GluK5^{-/-}$ where mossy fiber EPSCs cannot be detected in CA3 pyramidal neurons has challenged this view.[45] The authors argue that GluK5 is necessary for ionotropic signaling but not metabotropic signaling at mossy fiber synapses, thus contradicting previous results showing that GluK5 is indeed important for the metabotropic function of KARs.[28] At electron microscopy level, the sub-cellular distribution of KARs is altered in $GluK4^{-/-}$/$GluK5^{-/-}$ mice, where fewer immunogold particles for GluK2, 3 can be detected in postsynaptic densities despite that the overall level of expression of the protein remains similar.[45] Finally, functional studies in KAR subunit mutant mice have demonstrated that CA3 pyramidal neurons express GluK2, GluK4 and GluK5 subunits, consistent with specific immunohistochemical labelling in stratum lucidum where mossy fibers terminate.[28,43,45-49]

Action of KAR Agonists on the Slow After-Hyperpolarization in Hippocampal Pyramidal Neurons

The slow after-hyperpolarization following the action potential is a significant determinant of the firing pattern of peripheral and central neurons. Neurons accommodate their firing rate by means of different conductances activated upon depolarization and Ca^{2+} entry through L- and P/Q-type Ca^{2+} channels. Among channels participating in membrane hyperpolarization and burst termination, the family of Ca^{2+}-dependent K^+ channels plays a pivotal role in modulation of neuronal excitability (reviewed by refs. 50, 51). These channels mediate currents with distinct biophysical properties and sensitivity to pharmacological inhibitors, are coupled to 2nd messenger cascades and exhibit profound modulation by a spectrum of transmitters.[51] For instance, Type 1 metabotropic glutamate receptors (mGluRs) as well as cholinergic, monoaminergic, corticosteroid and VIP receptors, decrease I_{sAHP} in the cortex or sub-cortical areas of the brain.[52-59] The original finding that KARs modulate I_{sAHP} in central neurons is attributable to Melyan et al[26] in a report showing that exogenous kainate application persistently decreases I_{sAHP} in CA1 pyramidal neurons. This phenomenon was later reported in different cell types and in other hippocampal areas[24,28,45] indicating that kainate-induced modulation of I_{sAHP} might have wider adaptative significance. The GluK1 subunit does not seem to be involved in this effect in view of the poor pharmacological action of the GluK1-selective agonist ATPA as compared to kainate or domoate.[26] We found that bath application of kainate (50 nM) irreversibly decreased I_{sAHP} in CA3 pyramidal neurons (Fig. 1).[28] The effect of kainate on I_{sAHP} was dose dependent, with an EC_{50} of 6 nM. At this concentration,

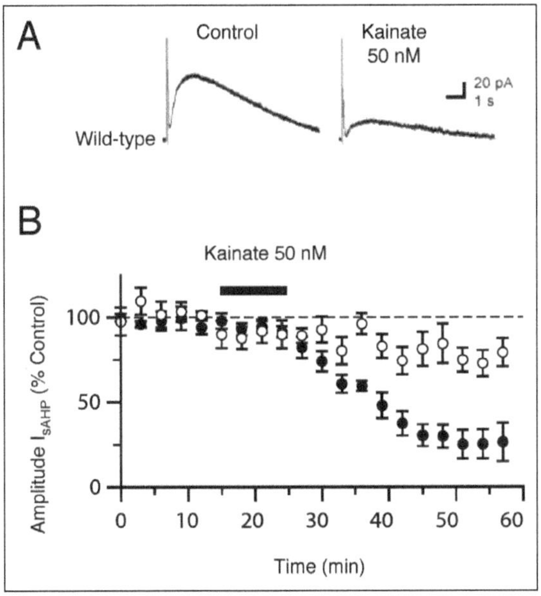

Figure 1. High-affinity KARs modulate I_{sAHP} in CA3 pyramidal neurons. A) Sample traces (average of 5 consecutive trials) showing the reduction of I_{sAHP} by bath application of kainate (50 nM). B) Plot of I_{sAHP} amplitude against time illustrating the reduction caused by kainate (filled circles) and the steady run-down observed when no treatment is given (open circles). Modified from: Ruiz A, Sachidhanandam S, Utvik JK et al. J Neurosci. 2005; 25:11710-11718.

kainate inhibited I_{sAHP} without significant variation in somatic holding current and input resistance of CA3 pyramidal neurons, consistent with a nonionotropic action. To identify the KAR subunits involved in I_{sAHP} modulation, we analyzed the effect of kainate in GluK2$^{-/-}$ and GluK5$^{-/-}$ mice. We observed no significant difference in I_{sAHP} characteristics obtained from both knock-out mice and wild-type littermate. Furthermore, kainate had no effect on I_{sAHP} amplitude in both GluK2$^{-/-}$ and GluK5$^{-/-}$ mice at a concentration up to 25 nM (Fig. 2). These results imply that in addition to GluK2, GluK5 subunits are required for the depression of I_{sAHP} observed with exogenous application of low nanomolar kainate concentration. This is in disagreement with the finding that inhibition of I_{sAHP} by low a concentration of kainate is intact in both GluK4$^{-/-}$ and GluK5$^{-/-}$ mice as well as in the double knockout GluK4$^{-/-}$/GluK5$^{-/-}$.[45] The reasons for this discrepancy are unclear but the latter report provides no information on I_{sAHP} characteristics in the different high-affinity KAR knockout mice, nor does it show the time course of the effect of kainate on I_{sAHP} corrected for experimentally occurring run-down of this current measured with the whole-cell configuration of the patch-clamp technique.

Synaptically Released Glutamate Modulates I_{sAHP} in Hippocampal Pyramidal Neurons

Whether endogenous glutamate alters cellular excitability by indirect regulation of ion channel function is central to this chapter. If progress has been made on the list of neurotransmitters and second messengers that can alter cellular excitability as a result of changes in I_{sAHP}, it is unclear how it occurs, notwithstanding the difficulty in linking pharmacological, biochemical and electrophysiological observations with the actions mediated by endogenous release from synaptic terminals. As stressed above, kainate irreversibly decreases I_{sAHP} elicited in CA1 and CA3 pyramidal neurons via G-proteins and subsequent PKC or PKA activation, suggesting that KARs may indeed signal through a metabotropic route that enhances cortical excitability. Furthermore, activity-dependent recruitment of metabotropic KARs by trains of stimuli applied to Schaffer collaterals mimicked the effect of kainate bath application, implying that metabotropic-like actions mediated by KARs can be triggered by synaptically released glutamate in CA1.[25-27] Because glutamatergic transmission at this pathway lacks a KAR-mediated component, KARs modulating I_{sAHP} might be different from those mediating KAR-EPSCs. This depression can also be elicited by Type 1 mGluRs and G-protein activation[58] and it has now become apparent that synaptically released glutamate can also modulate I_{sAHP} by acting on KARs in the CA3 area. We recently demonstrated that trains of stimuli designed to release glutamate at mossy fiber synapses depress I_{sAHP} in CA3 pyramidal neurons and that this depression is abolished in GluK5 knockout mice (Fig. 3). Synaptic modulation of I_{sAHP} was specific to the mossy fiber pathway, because no significant depression of I_{sAHP} amplitude was observed by activating perforant path synapses, or the associational/commissural pathway. Surprisingly, I_{sAHP} recorded in GluK2$^{-/-}$ mice was also insensitive to stimuli delivered to mossy fibers Similarly, the effect of stimulus trains on CA3 pyramidal cell firing recorded in current-clamp mode was absent in GluK5$^{-/-}$ mice indicating that GluK5 plays a critical role in the synaptic regulation of I_{sAHP} independently of the gating mechanisms responsible for KAR-EPSCs. These data were substantiated by immunohistochemical experiments in which we found that antibodies against GluK5 do not label stratum lucidum in GluK2$^{-/-}$ mice implying that GluK5 membrane expression is critically dependent on GluK2 (Fig. 4). Furthermore, using an anti-GluK2 antibody, a G-protein labeled by anti-$G_{\alpha q/11}$ antibody was co-immunoprecipitated with GluK2 and GluK5 subunits in wild-type and GluK1$^{-/-}$ but not GluK5$^{-/-}$ mice. These data argue that the lack of I_{sAHP} inhibition in GluK2$^{-/-}$ mice might be a consequence of the loss of GluK5 protein in CA3 pyramidal cells and is consistent with results showing that intracellular trafficking of GluK5 is regulated by endoplasmic reticulum retention signals which prevent the subunit expression at the plasma membrane in absence of one of the GluK1-GluK3 subunits.[60]

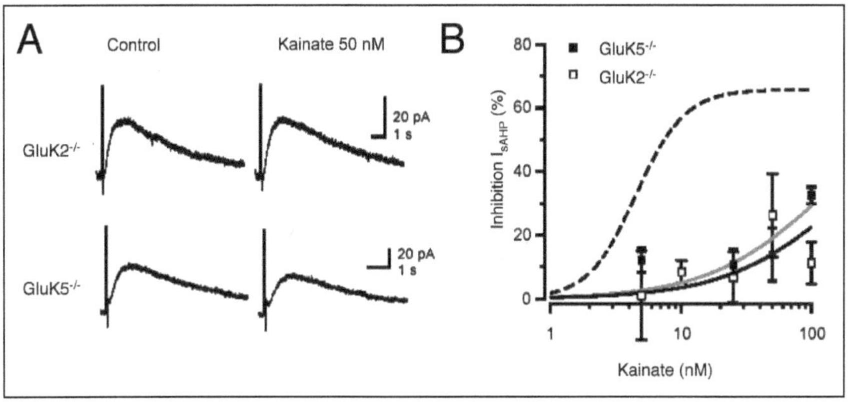

Figure 2. Kainate-induced modulation of I_{sAHP} is impaired in GluK2$^{-/-}$ and GluK5$^{-/-}$ mice. A) Traces (average of 5 consecutive trials) obtained in different KAR knockout mice showing that kainate (50 nM) had no effect on I_{sAHP} in GluK2$^{-/-}$ and GluK5$^{-/-}$ mice (B) Dose-response relationship for KA-induced reduction of I_{sAHP} in the different knockout. The curve obtained in the wild-type is represented by the dashed line for comparison. (A given KA concentration was tested in at least 3 cells; *n* = 21). Modified from: Ruiz A, Sachidhanandam S, Utvik JK et al. J Neurosci 2005; 25:11710-11718.

Candidate Mechanisms Linking KAR Activation and Enhanced Cellular Excitability

The findings presented above highlight that high-affinity KARs containing GluK5 subunits are good candidates for the expression of a metabotropic function in pyramidal neurons. Inhibition of I_{sAHP} by KARs shows a requirement for adenylate cyclase and PKA in CA1 pyramidal neurons. However, whilst KAR action is sensitive to PKA inhibitors, activation of PKC is sufficient to mimic the metabotropic function of KARs. The downstream effector seems to be the MAP kinase cascade which provides a bidirectional modulation of I_{sAHP}.[27] Furthermore, the requirement of G$\alpha_{i/o}$ proteins for metabotropic KAR-mediated actions is a consistent finding,[21,26-28,61,62] whereas the KAR itself represents an obvious candidate site for phosphorylation, most likely on GluK2.[63-65] KAR subunit deletion studies have shown that GluK2$^{-/-}$ or GluK5$^{-/-}$, but not GluK1$^{-/-}$ mice, lack kainate-induced inhibition of I_{sAHP}.[24,28] What is not yet understood is whether KARs couple to G-proteins directly, or if there is some involvement of ancillary proteins. In this regard, GluK2-containing receptors may be similar to AMPA, NMDA and nicotinic acetylcholine receptors, which also appear capable of activating second messenger systems.[53,66-68] The predicted sequence topology of GluK2-containing receptors is different from that of G protein-coupled receptors which normally have 7-transmembrane domains. We postulated the heretical idea that conformational changes caused by glutamate binding to GluK2 would allow the interaction with adaptor proteins (Fig. 5). This scenario might then be followed by G$\alpha_{i/o}$ protein activity which, in turn, would initiate a cascade of events involving protein-kinase activation and finally the modulation of channels responsible for I_{sAHP} in CA3 pyramidal neurons. Whether a single KAR complex can initiate both forms of signaling via distinct subunits however remains to be demonstrated.

Conclusion

The slow Ca^{2+} activated K^+ current I_{sAHP} which follows a train of action potentials is a privileged target for modulation of cellular excitability by transmitter receptors.[69] Several

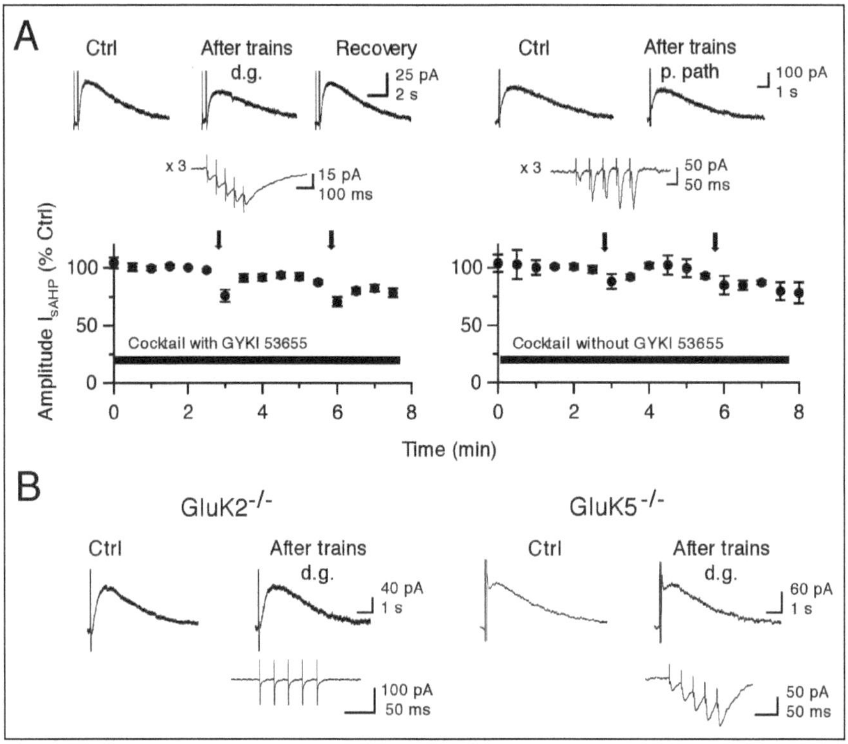

Figure 3. Synaptically-released glutamate inhibits I_{sAHP} via high-affinity KARs. A) Left, time course of I_{sAHP} amplitude showing a reversible decrease after 20 Hz tetanic stimulation in the dentate granule cell layer (train d.g.) at the time indicated by the arrows. Right, A nonsignificant decrease was observed after lateral perforant pathway (p. path) stimulation. Top sample traces show I_{sAHP} taken before (average of 3 traces), 350 ms after the first series of trains (single trace) and the recovery. Bottom traces show 5 consecutive KAR-EPSCs evoked by the train of stimuli. B) Top, sample traces showing I_{sAHP} recorded from GluK2$^{-/-}$ and GluK5$^{-/-}$ mice. I_{sAHP} was unaffected by induction of the stimulus protocol in GluK2$^{-/-}$ or GluK5$^{-/-}$ mice. Bottom, Trains of EPSCs displayed facilitation and disappeared during application of GYKI 53655 in GluK2$^{-/-}$ mice. In GluK5$^{-/-}$ mice, the trains of stimuli elicited KAR-EPSCs with marked summation typical of mossy fiber responses. Modified from: Ruiz A, Sachidhanandam S, Utvik JK et al. J Neurosci 2005; 25:11710-11718.

reports including our own have shown that exogenous application of KAR agonists inhibit I_{sAHP} in CA3 pyramidal neurons and that endogenous glutamate mimics this depression. However, the identity of the KAR subunits involved in this effect, as well as the fine details of the molecular steps leading to I_{sAHP} modulation, are still under extensive scrutiny. Interestingly, a parallel can be drawn between KARs and mGluRs whereby high-affinity glutamate receptors increase neuronal firing by modulating a subclass of K^+ channels. During development, other K^+ channels sensitive to the bee neurotoxin apamin which are responsible for the medium AHP are linked to KAR activation and modulate the firing of hippocampal interneurons.[70] The functional ramifications of these findings emphasize the diversity and the complexity of KAR-mediated actions—a 'boulevard' for future investigations on the fundamental basis of cortical network excitability.

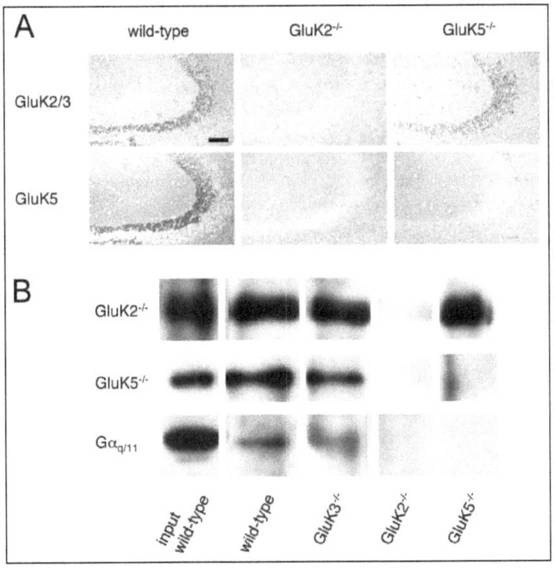

Figure 4. The GluK5 subunit as putative candidate for the modulation of I_{sAHP} in CA3 pyramidal neurons. A) Immunostaining of hippocampal sections from wild-type, GluK2$^{-/-}$ and GluK5$^{-/-}$ mice with anti- GluK2/3 and anti-GluK5 antibodies. Both antibodies labeled stratum lucidum but not stratum radiatum. GluK5 immunoreactivity was lost in the CA3 region of GluK2$^{-/-}$ mice. B) Western blots showing that anti-GluK2 antibodies co-immunoprecipitate both GluK5 and GluK2 subunits in wild-type and GluK1$^{-/-}$ mice but not in GluK5$^{-/-}$ mice. $G\alpha_{q/11}$ was also immunoprecipitated under the same conditions. Modified from: Ruiz A, Sachidhanandam S, Utvik JK et al. J Neurosci 2005; 25:11710-11718.

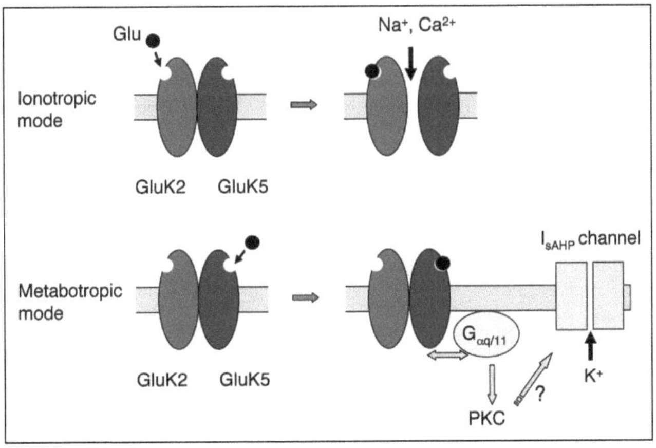

Figure 5. A speculative model for KARs with a bi-modal function in CA3 pyramidal neurons. A single heteromeric receptor operates with simultaneous double signaling through binding of glutamate to two categories of sites, leading to the opening of an ion channel by glutamate binding to the GluK2 subunit (Top) and a metabotropic effect causing I_{sAHP} inhibition via G-protein activation upon glutamate binding on the GluK5 subunit (Bottom). Modified from: Ruiz A, Sachidhanandam S, Utvik JK et al. J Neurosci 2005; 25:11710-11718.

References

1. Agrawal SG, Evans RH. The primary afferent depolarizing action of kainate in the rat. Br J Pharmacol 1986; 87:345-355.
2. Bleakman D, Ballyk BA, Schoepp DD et al. Activity of 2, 3-benzodiazepines at native rat and recombinant human glutamate receptors in vitro: stereospecificity and selectivity profiles. Neuropharmacology 1996; 35:1689-1702.
3. Engberg I, Tarnawa I, Durand J et al. An analysis of synaptic transmission to motoneurones in the cat spinal cord using a new selective receptor blocker. Acta Physiol Scand 1993; 148:97-100.
4. Ruiz A, Durand J. Modulation of kainate-induced responses by pentobarbitone and GYKI-53784 in rat abducens motoneurons in vivo. Brain Res 1999; 818:421-430.
5. Castillo PE, Malenka RC, Nicoll RA. Kainate receptors mediate a slow postsynaptic current in hippocampal CA3 neurons. Nature 1997; 388:182-186.
6. Cossart R, Esclapez M, Hirsch JC et al. GluR5 kainate receptor activation in interneurons increases tonic inhibition of pyramidal cells. Nat Neurosci 1998; 1:470-478.
7. DeVries SH, Schwartz EA. Kainate receptors mediate synaptic transmission between cones and 'Off' bipolar cells in a mammalian retina. Nature 1999; 397:157-160.
8. Frerking M, Malenka RC, Nicoll RA. Synaptic activation of kainate receptors on hippocampal interneurons. Nat Neurosci 1998; 1:479-486.
9. Kidd FL, Isaac JT. Developmental and activity-dependent regulation of kainate receptors at thalamocortical synapses. Nature 1999; 400:569-573.
10. Li P, Wilding TJ, Kim SJ et al. Kainate-receptor-mediated sensory synaptic transmission in mammalian spinal cord. Nature 1999; 397:161-164.
11. Vignes M, Collingridge GL. The synaptic activation of kainate receptors. Nature 1997; 388:179-182.
12. Sachidhanandam S, Blanchet C, Jeantet Y et al. Kainate receptors act as conditional amplifiers of spike transmission at hippocampal mossy fiber synapses. J Neurosci 2009; 29:5000-5008.
13. Barberis A, Sachidhanandam S, Mulle C. GluR6/KA2 kainate receptors mediate slow-deactivating currents. J Neurosci 2008; 28:6402-6406.
14. Perrais D, Coussen F, Mulle C. Atypical functional properties of GluK3-containing kainate receptors. J Neurosci 2009; 29:15499-15510.
15. Chittajallu R, Vignes M, Dev KK et al. Regulation of glutamate release by presynaptic kainate receptors in the hippocampus. Nature 1996; 379:78-81.
16. Clarke VR, Ballyk BA, Hoo KH et al. A hippocampal GluR5 kainate receptor regulating inhibitory synaptic transmission. Nature 1997; 389:599-603.
17. Contractor A, Swanson GT, Sailer A et al. Identification of the kainate receptor subunits underlying modulation of excitatory synaptic trnsmission in the CA3 region of the hippocampus. J Neurosci 2000; 20:8269-8278.
18. Kamiya H, Ozawa S. Kainate receptor-mediated presynaptic inhibition at the mouse hippocampal mossy fibre synapse. J Physiol (Lond) 2000; 523 Pt 3:653-665.
19. Kidd FL, Coumis U, Collingridge GL et al. A presynaptic kainate receptor is involved in regulating the dynamic properties of thalamocortical synapses during development. Neuron 2002; 34:635-646.
20. Lauri SE, Delany C, VR JC et al. Synaptic activation of a presynaptic kainate receptor facilitates AMPA receptor-mediated synaptic transmission at hippocampal mossy fibre synapses. Neuropharmacology 2001; 41:907-915.
21. Rodriguez-Moreno A, Herreras O, Lerma J. Kainate receptors presynaptically downregulate GABAergic inhibition in the rat hippocampus. Neuron 1997; 19:893-901.
22. Schmitz D, Frerking M, Nicoll RA. Synaptic activation of presynaptic kainate receptors on hippocampal mossy fiber synapses. Neuron 2000; 27:327-338.
23. Schmitz D, Mellor J, Nicoll RA. Presynaptic kainate receptor mediation of frequency facilitation at hippocampal mossy fiber synapses. Science 2001; 291:1972-1976.
24. Fisahn A, Heinemann SF, McBain CJ. The kainate receptor subunit GluR6 mediates metabotropic regulation of the slow and medium AHP currents in mouse hippocampal neurones. J Physiol 2005; 562(Pt 1):199-203.
25. Melyan Z, Lancaster B, Wheal HV. Metabotropic Regulation of Intrinsic Excitability by Synaptic Activation of Kainate Receptors. The Journal of Neuroscience 2004; 24:4530-4534.
26. Melyan Z, Wheal HV, Lancaster B. Metabotropic-mediated kainate receptor regulation of IsAHP and excitability in pyramidal cells. Neuron 2002; 34:107-114.
27. Grabauskas G, Lancaster B, O'Connor V et al. Protein kinase signalling requirements for metabotropic action of kainate receptors in rat CA1 pyramidal neurones. J Physiol 2007; 579(Pt 2):363-373.
28. Ruiz A, Sachidhanandam S, Utvik JK et al. Distinct subunits in heteromeric kainate receptors mediate ionotropic and metabotropic function at hippocampal mossy fiber synapses. J Neurosci 2005; 25:11710-11718.

29. Bowie D. Ion-dependent gating of kainate receptors. J Physiol 2010; 588(Pt 1):67-81.
30. Rodriguez-Moreno A, Sihra TS. Kainate receptors with a metabotropic modus operandi. Trends Neurosci 2007; 30:630-637.
31. Coussen F, Mulle C. Kainate receptor-interacting proteins and membrane trafficking. Biochem Soc Trans 2006; 34(Pt 5):927-930.
32. Alt A, Weiss B, Ogden AM et al. Pharmacological characterization of glutamatergic agonists and antagonists at recombinant human homomeric and heteromeric kainate receptors in vitro. Neuropharmacology 2004; 46:793-806.
33. Jatzke C, Watanabe J, Wollmuth LP. Voltage and concentration dependence of Ca($^{2+}$) permeability in recombinant glutamate receptor subtypes. J Physiol 2002; 538(Pt 1):25-39.
34. Paternain AV, Rodriguez-Moreno A, Villarroel A et al. Activation and desensitization properties of native and recombinant kainate receptors. Neuropharmacology 1998; 37(10-11):1249-1259.
35. Wenthold RJ, Trumpy VA, Zhu WS et al. Biochemical and assembly properties of GluR6 and KA2, two members of the kainate receptor family, determined with subunit-specific antibodies. J Biol Chem 1994; 269:1332-1339.
36. Wisden W, Seeburg PH. A complex mosaic of high-affinity kainate receptors in rat brain. J Neurosci 1993; 13:3582-3598.
37. Werner P, Voigt M, Keinanen K et al. Cloning of a putative high-affinity kainate receptor expressed predominantly in hippocampal CA3 cells. Nature 1991; 351:742-744.
38. Pinheiro PS, Perrais D, Coussen F et al. GluR7 is an essential subunit of presynaptic kainate autoreceptors at hippocampal mossy fiber synapses. Proc Natl Acad Sci USA 2007; 104:12181-12186.
39. Lerma J, Paternain AV, Naranjo JR et al. Functional kainate-selective glutamate receptors in cultured hippocampal neurons. Proc Natl Acad Sci USA 1993; 90:11688-11692.
40. Perrais D, Pinheiro PS, Jane DE et al. Antagonism of recombinant and native GluK3-containing kainate receptors. Neuropharmacology 2009; 56:131-140.
41. Herb A, Higuchi M, Sprengel R et al. Q/R site editing in kainate receptor GluR5 and GluR6 premRNAs requires distant intronic sequences. Proc Natl Acad Sci USA 1996; 93:1875-1880.
42. Marchal C, Mulle C. Postnatal maturation of mossy fibre excitatory transmission in mouse CA3 pyramidal cells: a potential role for kainate receptors. J Physiol 2004; 561(Pt 1):27-37.
43. Mulle C, Sailer A, Perez-Otano I et al. Altered synaptic physiology and reduced susceptibility to kainate-induced seizures in GluR6-deficient mice. Nature 1998; 392:601-605.
44. Contractor A, Sailer AW, Darstein M et al. Loss of kainate receptor-mediated heterosynaptic facilitation of mossy-fiber synapses in KA2-/- mice. J Neurosci 2003; 23:422-429.
45. Fernandes HB, Catches JS, Petralia RS et al. High-affinity kainate receptor subunits are necessary for ionotropic but not metabotropic signaling. Neuron 2009; 63:818-829.
46. Darstein M, Petralia RS, Swanson GT et al. Distribution of kainate receptor subunits at hippocampal mossy fiber synapses. J Neurosci 2003; 23:8013-8019.
47. Bureau I, Bischoff S, Heinemann SF et al. Kainate receptor-mediated responses in the CA1 field of wild-type and GluR6-deficient mice. J Neurosci 1999; 19:653-663.
48. Petralia RS, Wang YX, Wenthold RJ. Histological and ultrastructural localization of the kainate receptor subunits, KA2 and GluR6/7, in the rat nervous system using selective antipeptide antibodies. J Comp Neurol 1994; 349:85-110.
49. Fogarty DJ, Perez-Cerda F, Matute C. KA1-like kainate receptor subunit immunoreactivity in neurons and glia using a novel anti-peptide antibody. Brain Res Mol Brain Res 2000; 81(1-2):164-176.
50. Sah P, Faber ES. Channels underlying neuronal calcium-activated potassium currents. Prog Neurobiol 2002; 66:345-353.
51. Vogalis F, Storm JF, Lancaster B. SK channels and the varieties of slow after-hyperpolarizations in neurons. Eur J Neurosci 2003; 18:3155-3166.
52. Young SR, Chuang SC, Wong RK. Modulation of afterpotentials and firing pattern in guinea pig CA3 neurones by group I metabotropic glutamate receptors. J Physiol 2004; 554(Pt 2):371-385.
53. Krause M, Offermanns S, Stocker M et al. Functional specificity of G alpha q and G alpha 11 in the cholinergic and glutamatergic modulation of potassium currents and excitability in hippocampal neurons. J Neurosci 2002; 22:666-673.
54. Scroggs RS, Cardenas CG, Whittaker JA et al. Muscarine reduces calcium-dependent electrical activity in substantia nigra dopaminergic neurons. J Neurophysiol 2001; 86:2966-2972.
55. Haug T, Storm JF. Protein kinase A mediates the modulation of the slow Ca($^{2+}$)-dependent K($^{+}$) current, I(sAHP), by the neuropeptides CRF, VIP and CGRP in hippocampal pyramidal neurons. J Neurophysiol 2000; 83:2071-2079.
56. Tegner J, Grillner S. Interactive effects of the GABABergic modulation of calcium channels and calcium-dependent potassium channels in lamprey. J Neurophysiol 1999; 81:1318-1329.

57. Carrer HF, Araque A, Buno W. Estradiol regulates the slow Ca^{2+}-activated K^+ current in hippocampal pyramidal neurons. J Neurosci 2003; 23:6338-6344.
58. Heuss C, Scanziani M, Gahwiler BH et al. G-protein-independent signaling mediated by metabotropic glutamate receptors. Nat Neurosci 1999; 2:1070-1077.
59. Pockett S. Dopamine changes the shape of action potentials in hippocampal pyramidal cells. Brain Res 1985; 342:386-390.
60. Ren Z, Riley NJ, Garcia EP et al. Multiple trafficking signals regulate kainate receptor KA2 subunit surface expression. J Neurosci 2003; 23:6608-6616.
61. Frerking M, Schmitz D, Zhou Q et al. Kainate receptors depress excitatory synaptic transmission at CA3—>CA1 synapses in the hippocampus via a direct presynaptic action. J Neurosci 2001; 21:2958-2966.
62. Rodriguez-Moreno A, Lerma J. Kainate receptor modulation of GABA release involves a metabotropic function. Neuron 1998; 20:1211-1218.
63. Dildy-Mayfield JE, Harris RA. Activation of protein kinase C inhibits kainate-induced currents in oocytes expressing glutamate receptor subunits. J Neurochem 1994; 62:1639-1642.
64. Hirbec H, Francis JC, Lauri SE et al. Rapid and differential regulation of AMPA and kainate receptors at hippocampal mossy fibre synapses by PICK1 and GRIP. Neuron 2003; 37:625-638.
65. Raymond LA, Blackstone CD, Huganir RL. Phosphorylation and modulation of recombinant GluR6 glutamate receptors by cAMP-dependent protein kinase. Nature 1993; 361:637-641.
66. Callister RJ, Keast JR, Sah P. $Ca(^{2+})$-activated K^+ channels in rat otic ganglion cells: role of Ca^{2+} entry via Ca^{2+} channels and nicotinic receptors. J Physiol 1997; 500 (Pt 3):571-582.
67. Cox CL, Metherate R, Ashe JH. Modulation of cellular excitability in neocortex: muscarinic receptor and second messenger-mediated actions of acetylcholine. Synapse 1994; 16:123-136.
68. Pedarzani P, Storm JF. Evidence that Ca/calmodulin-dependent protein kinase mediates the modulation of the Ca^{2+}-dependent K^+ current, IAHP, by acetylcholine, but not by glutamate, in hippocampal neurons. Pflugers Arch 1996; 431:723-728.
69. Madison DV, Nicoll RA. Control of the repetitive discharge of rat CA 1 pyramidal neurones in vitro. J Physiol 1984; 354:319-331.
70. Segerstrale M, Juuri J, Lanore F et al. High firing rate of neonatal hippocampal interneurons is caused by attenuation of afterhyperpolarizing potassium currents by tonically active kainate receptors. J Neurosci 2010; 30:6507-6514.

Chapter 7

Metabotropic Actions of Kainate Receptors in Dorsal Root Ganglion Cells

Jose Luis Rozas*

Abstract

Kainate receptors are widely distributed in the CNS, but also in the PNS. Dorsal root ganglia are enriched in this subtype of glutamate ionotropic receptors. In addition to their activity as ligand-gated ion channels, kainate receptors exhibit other properties already characterized in other systems, such as hippocampus, i.e., their ability to induce a metabotropic cascade signalling, through G-protein and PKC activation. With a very similar actuation mechanism as formerly described in the CNS, kainate receptors in the DRG also present other differentiated features, such as the Ca^{2+} channel blockade and a self-regulation property. The peculiarity of these neurons has served to progress the study of kainate receptors. Nevertheless, many other physiological functions of these receptors remain unclear, as does the related molecular nature of the metabotropic cascade and the involvement of this signalling pathway with sensory transmission of pain.

Introduction

Dorsal root ganglia (DRGs) contain the cell bodies of primary sensory neurons. These neurons carry somatosensorial information from the body to the central nervous system (CNS) and have a bipolar morphology with an undifferentiated terminal at the periphery and synaptic terminal entering the dorsal horn of the spinal cord. Their synapses are glutamatergic although they can also release neuropeptides, such as substance P.

In this chapter, we focused on a specific type of DRG sensory neuron, i.e., neurons carrying nociceptive information. Nociceptive fibers that penetrate into the spinal cord come from DRGs belonging to two classes: Aδ and C fibers. The first type are myelinated fibers ("fast" conducting fibers) carrying mechanosensitive and thermal information. The second are "slow" fibers which are unmyelinated and carry information elicited by pain stimuli and temperature changes. Despite of the number of receptors expressed by these neurons (for several typical examples, see refs. 1-4), they operate as primary afferents and do not receive synapses from any other neuron. In cocultures of DRG and spinal cord neurons, DRG neurons make synaptic contacts with dorsal horn neurons. This property is advantageous to clearly separate presynaptic from postsynaptic effects.

DRG neurons express glutamate receptors, including those of the kainate type. Kainate receptors are broadly distributed in both the CNS and the PNS. Agrawal and Evans[5] demonstrated that kainate application induced depolarization at the DRG of immature rats, as well as a reversible

*Jose Luis Rozas—Instituto de Biomedicina de Sevilla (IBIS), Hospital Universitario Virgen del Rocío/CSIC/Universidad de Sevilla and Dpto. Fisiología Médica y Biofísica, Universidad de Sevilla and Centro de Investigación Biomédica en Red sobre Enfermedades Neurodegenerativas (CIBERNED), Sevilla, Spain. Email: jlrozas@us.es

Kainate Receptors: Novel Signaling Insights, edited by Antonio Rodríguez-Moreno and Talvinder S. Sihra.

depression of the afferent volley. Others ionotropic glutamate receptor agonists, like domoate or 5-Br-willardiine (5-IW) caused similar effects. These actions were located on C fibers, something that was later corroborated by immunocytochemical methods.[6]

Native kainate receptors were initially identified in DRG cultured neurons in patch clamp experiments.[7] Electrophysiological responses were evoked by agonists like domoate or quisqualate, but not by NMDA. This type of responses showed long-lasting desensitization to kainate and glutamate which was unusual compared to any other central glutamate receptor previously analyzed. Responses induced by glutamate or quisqualate suffered complete desensitization, while the desensitization mediated by kainate or domoate caused an initial peak followed by a slow decay before reaching a steady-state.

Finally, the development of pharmacological tools and cloning of kainate receptor subunits,[8-12] greatly advanced the characterization of both peripheral and CNS receptors. The initial availability of 2-3 benzodiazepines like GYKI 53655 allowed definitive discrimination of AMPA and kainate receptors.[13,14] Other antagonists have allowed more recent discrimination of receptors with different subunit composition.[15,16] The generation of knockout mice lacking kainate receptors has significantly contributed to the understanding of the biology of the kainate receptors,[17-20] although the rising vision is not exempt of new and exciting questions.

Expression and Subunit Composition

The study of the composition of native kainate receptors has turned out to be less controversial at the DRG neurons than at the CNS, where, particularly at the hippocampus, there is a mosaic of receptors expressed depending on the neuronal type.[21-24] Apparently, this complexity does not occur in DRG neurons, where GluR5 is the most abundantly expressed. Although mRNA for GluR6 and GluR7 have also detected, it is unlikely that they contribute to form functional receptors since the kainate-induced current are completely lacking in GluR5 deficient animals.[25] DRG neurons also express KA1 and KA2, the subunits of high affinity for kainate, that may contribute to assemble functional heteromeric receptors together with GluR5.

The development of genetic and molecular tools has helped vastly to figure out the composition of kainate receptors in DRG neurons. The characterization of mice lacking GluR5 subunits clearly showed the complete absence of currents in DRG dissociated neurons in response to kainate.[26] This also excludes the possibility that kainate induces responses through AMPA receptors, although some studies have indicated the existence of a subpopulation of DRG where functional AMPA receptor could be expressed.[27,28] On the other hand, the required association of KA1 or KA2 subunits with the other three low-affinity subunits (GluR5-7) to assemble functional receptors, due to the impossibility of KA1 and KA2 for making functional receptors,[12,29] makes it difficult to determine their specific roles. However, the functional analysis in heterologous systems, such as HEK cells, expressing specific subunits indicates that native DRG responses are mostly reproduced by GluR5/KA2 since the GluR5/KA2 desensitization mimics responses obtained in DRG neurons and they do not present the fast desensitization typically observed from GluR6/KA1 or GluR6/KA2 combinations.[25,29] In rat trigeminal ganglia, GluR5/KA2 is the preferential composition of kainate receptors[30] and it is possible that this combination is the most widely extended in DRGs.

The avalaibility of new pharmacological compounds has contributed notably to the knowledge of receptor composition in DRG neurons (for a detailed review on the agonists and antagonists of kainate receptors in DRG neurons and other systems, see Lerma et al 2001; Lerma, 2006 and Pinheiro and Mulle, 2006[31-33]).The sensitivity to ATPA, the specific agonist for GluR5-containing kainate receptors, corroborates, once again, that this subunit is a fundamental part of the kainate receptor in DRGs (EC50 = 0.6-1.3 μM[34]).Domoate and 5-IW, are also more potent agonists for GluR5, exhibiting a low EC50 in DRG neurons. A low IC50 for the specific antagonist of the subunit GluR5, LY382884 (around 1 μM) shows that this subunit is the principal one in DRG neurons. In contrast, the same drug fails to inhibit homomeric GluR6 receptors, but not heteromeric GluR5/GluR6, recombinantly expressed in heterologous systems.[16] Therefore, it seems clearly established that GluR5 is a mandatory subunit to get functional kainate receptors assembled in DRG neurons. Importantly though, the effectiveness of these agonists is higher for native DRG receptors than for GluR5 homomeric recombinant receptors, indicating that there are additional subunits composing the receptor. Alternatively, the existence of unknown interacting proteins affecting GluR5 properties cannot be refuted at this stage.[35,36]

Specific antibodies against kainate receptors, particularly against the GluR5 subunit, have been difficult to raise, since the only avalaible antibody against this subunit was unable to differentiate between GluR5, 6 and 7 subunits.[37] Later, however, the use of some new antibodies have confirmed the predominance of GluR5 in native kainate receptors in DRG neurons.[38] Immunofluorescence labelling has demonstrated the presence of GluR5, KA1 and KA2 at the synaptic terminals of DRG in the dorsal horn. Presynaptic staining of GluR5 is restricted to laminae I-III of the dorsal horn. However, GluR5 is also located at the postsynaptic sites forming receptors with KA1-KA2 or even with GluR6 subunits.[6]

Postranscriptional Processing

In this chapter, I focus on GluR5 variants, since GluR6 and GluR7 subunits are almost, if not completely absent in DRG neurons. Four splice variants of the GluR5 have been reported, named GluR5a-d. The GluR5a variant possesses the shortest cytoplasmic C-terminal domain, extending only 16 amino-acid downstream from the third transmembrane segment (from position 854). GluR5b possesses, besides those 16 residues, a tail of 49 additional amino-acids.[8] GluR5c presents 29 amino-acids insert before the 49 residues tail of GluR5b.[10] The GluR5d variant shows a completely different 49 amino-acid tail at 854 position.[11] At the dorsal horn of the spinal cord, the subunit preferentially expressed is GluR5a. However, for DRG neurons, the specific splice variant is currently unknown and it is not clear if there is a predominant splice variant. Recent studies have identified an important role for specific sequences (retention signals) of splice variants of kainate receptors. Besides controlling the subunit retention at the endoplasmic reticulum,[39] specific sequences can play a key role in the signalling mediated by the receptors and in phosphorylation mediated regulation if the spliced inserts bearing phosphorylation sites.[40] Details on regulation will be discussed below in the "Kainate receptors plasticity" section.

The mRNA editing could be considered as a biological strategy to change biophysical properties of the edited receptor and its influence on neuronal physiology. Q to R editing (Q/R) occurs during mRNA maturation as a codon modification resulting from the change of a glutamine (Q636) to an arginine (R636).[41,42] This modification causes substantial changes in the biophysics of the ion channel: it reduces the unitary conductance, Ca^{2+}-permeability and the I/V curve changes from inward-rectifier to outward rectification (rectification index ≥ 1).[43,44]

The editing process seems to be regulated during development and maturation. During the embryonic period, the level of edited GluR5 mRNA is hardly detectable. Nevertheless, a few days later it reaches 50% of total GluR5 mRNA[45] in brain tissue. DRG are not an exception. In DRG neurons cultured from E18, only around 20% of mRNA is edited, consistent with the remarkable permeability to Ca^{2+} of kainate receptors at this stage. In contrast, in cultured neurons extracted from P7 mice, the edition level reach 97%, leading to a reduced conductance and almost null capacity to permeate Ca^{2+}.[46]

The biological significance of the kainate receptors editing, particularly of the GluR5, subunit has not been fully clarified yet. The temporary pattern of editing of kainate receptors coincides with the maturation timing of synaptic contacts of C fibers at the dorsal horn. Thus, it has been suggested that Ca^{2+}-permeable kainate receptors are important in the development and maturation of those synapses.[46] However, in generated transgenic mice whose receptors are permanently edited and present a six-fold reduction in kainate receptors dependent currents at DRG neurons,[47] development and maturation of synapses in the spinal cord is normal. In addition, these mice do not show any alteration in the response to chemical or thermal painful stimuli. It is therefore probable that kainate receptor edition plays a fine regulatory role in fiber physiology rather than as an essential general mechanism in synaptic development/maturation.

Kainate Receptors Physiology and Plasticity at Dorsal Root Ganglion Neurons

Synaptic transmission between DRG and dorsal horn neurons of the spinal cord depends on all three types of well-known ionotropic glutamate receptors: NMDA, AMPA and kainate. Kainate receptors mediate sensory synaptic transmission at the spinal cord, where they accomplish a double task: when being expressed by the neurons of the dorsal horn, they regulate the excitability of the postsynaptic membrane and given their localization at the presynaptic terminal of DRG fibers they modulate glutamate release.[48,49]

Activation of kainate receptors at the presynaptic terminal tends to decrease the amplitude of EPSC recorded in the dorsal horn. Using cocultures of dorsal horn and DRG neurons, Kerchner and collaborators[26] found that GluR5 subunits are essential for assembling ionotropic kainate receptors in DRG in that the modulation of the synaptic transmission is sensitive to ATPA. Indeed, in cocultured cells from GluR5 knock-out mice, this modulation by kainate is totally abolished. At the dorsal horn, however, both GluR5 and GluR6 contribute to assemble functional receptors,[a] as demonstrated by the fact that ionotropic responses are still present in dorsal horn from GluR5 or GluR6 knock-out mice.

Previously, Kerchner and collaborators also showed a modulation of GABA and glycine release from dorsal horn interneurons mediated by kainate receptors.[50,51] In this case, the modulation was bidirectional, since facilitation or inhibition of GABA release depended on the degree of kainate receptor activation. The proposed mechanism for this phenomenon is compatible with their ionotropic pathway. In principle, the activation of kainate receptors located at the interneuron presynaptic terminal would be able to depolarize the terminal favouring the Ca^{2+} influx and therefore the release of GABA/glycine. An excess of GABA would activate receptors located in the same terminal and would then inhibit its own release (see Fig. 1).

Is this mechanism of presynaptic modulation induced by kainate receptors similar to that found at the DRG terminals? In principle, it has not been possible to corroborate the facilitation induced by kainate receptors at DRG neurons, as would be expected for the ionotropic depolarization and subsequent regulation of glutamate release. Surprisingly, this mechanism of modulation in DRG terminals seems to be more in line with the mechanism described at hippocampal inhibitory synapses,[52,53] where a second messenger mediated signalling cascade plays a part.

Metabotropic Activity of Kainate Receptors

Kainate receptors present in DRG neurons activate a signalling cascade. This signalling pathway is not expected for ionotropic receptors. Rather, on the contrary, it is in keeping with and has similarities with the pathways activated by conventional metabotropic glutamate receptors. The use of Ca^{2+} imaging techniques was crucial to reveal some aspects of this particular behaviour of kainate receptors.[54]

This signalling pathway includes the activation of a Pertussis toxin sensitive G-protein and subsequently two differentiated steps: the release of Ca^{2+} from intracellular stores, presumably due to the production of IP3 and the activation of a PKC that leads to the inhibition of voltage dependent Ca^{2+} channels (VGCC) (see Fig. 2). This signal cascade activated by kainate receptors was previously described in hippocampal inhibitory terminals,[53,55] which express GluR5 subunit, but also in excitatory terminals[56] and in postsynaptic sites.[57]

The Ca^{2+} release from intracellular stores is a well-known and widely studied phenomenon[58,59] (for detailed review see Berridge et al, 2003; Clapham 2007[60,61]).The implications of Ca^{2+} as a secondary messenger in intracellular signalling as well as the influence that it could have on the exocytosis of neurotransmitter process have also been intensely studied.[62,63] Although these data provided direct evidence of the metabotropic activity of kainate receptors, at the moment, many details on the mechanisms of kainate-induced Ca^{2+} release are still poorly understood. For example,

[a] DRG neurons from GluR7 KO mice have not been analyzed. It is possible that this subunit contributes, along with others low-affinity subunits, to form kainate receptors at the postsynaptic site.

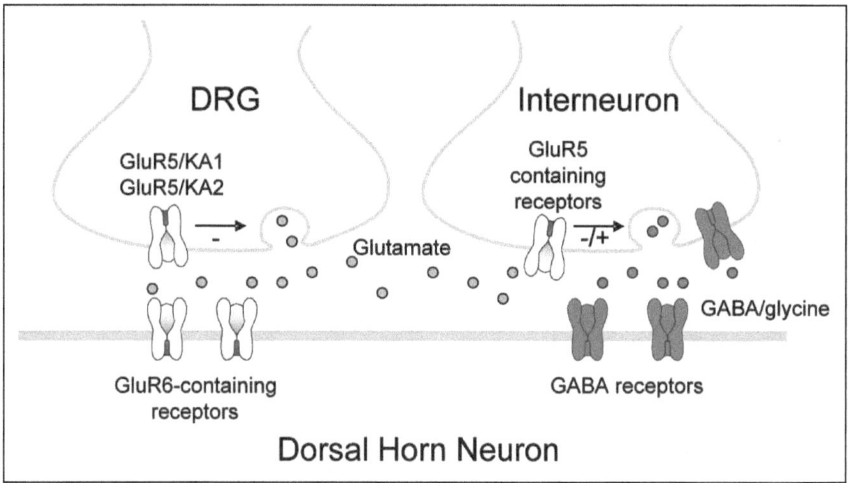

Figure 1. Kainate receptor regulates the glutamate and GABA/glycine release at the dorsal horn. At the DRG nerve terminal, GluR5 heteromers negatively regulates the glutamate release. The neurotransmitter activates AMPA, NMDA and kainate receptors (GluR6-containing). If the amount of glutamate is high, neurotransmitter spillover could activates presynaptic kainate receptors at medullar interneurons terminals (GluR5 composed). The depolarization induced by receptor activation produce the release of GABA/glycine, but these neurotransmitter could finally activate presynaptic GABA receptors, leading to the inhibition of release. (Adapted from Kerchner et al Neuron 2001; 32:477-88.)

we do not know which reservoir, pathway or intracellular receptors are involved in Ca^{2+} release or if there is IP3 production or another similar messenger. Nevertheless, the activation of the kainate receptor is able to induce the increase of intracellular Ca^{2+} even in the absence of extracellular Ca^{2+} and it is sensitive to the previous emptying of these deposits. Interestingly, the release of Ca^{2+} is confined to discrete membrane patches in DRG cells neurites (where kainate receptors are located). To date, this activity triggered by kainate receptor activation has not been studied in central neurons, although kainate receptors have been associated with Ca^{2+} induced-Ca^{2+} release (a process that is not initiated by a metabotropic cascade) in the mossy fibers of the hippocampus.[64] The function of the cytosolic Ca^{2+} increase in DRG needs further investigation.

The activation of PKC, in some of its multiple isoforms, by diacylglycerol (DAG) or phorbol esters, has been considered as a positive regulator of neurotransmitter release in central neurons.[65] Nevertheless, recent studies revealed that Munc13, also activated by DAG, can be directly responsible for this phenomenon[66]; this field is currently matter of intensive discussion.[67] However, in sensory neurons, PKC activation exerts the opposite effect, since phosphorylation of N-type calcium channels by PKC results in a decrease of their conductance,[68-70] especially in a subset of neurons which express syntaxin 1A.[71] This inhibition is of different nature to the inhibition produced by direct interaction of Ca^{2+} channels with the G-protein $\beta\gamma$ subunits.[72] The channel phosphorylation produces its voltage-independent inhibition, whereas $\beta\gamma$ interaction could be removed by strong depolarizing prepulses. As a consequence, the inhibition of VGCC induced by kainate results in a lower $[Ca^{2+}]$ in the presynaptic terminal after the arrival of the action potential and the subsequent decrease in glutamate release.

It is necessary to point out two significant aspects about this phenomenon: (1) the inhibitory action of kainate receptors activation on the Ca^{2+} influx is independent of the ionic activity of the receptor channel. The opening of the receptor causes the influx of Na^+ and Ca^{2+} ions through the channel to the cytosol. This Ca^{2+} can inhibit VGCC,[73] while the increase in Na^+ conductance

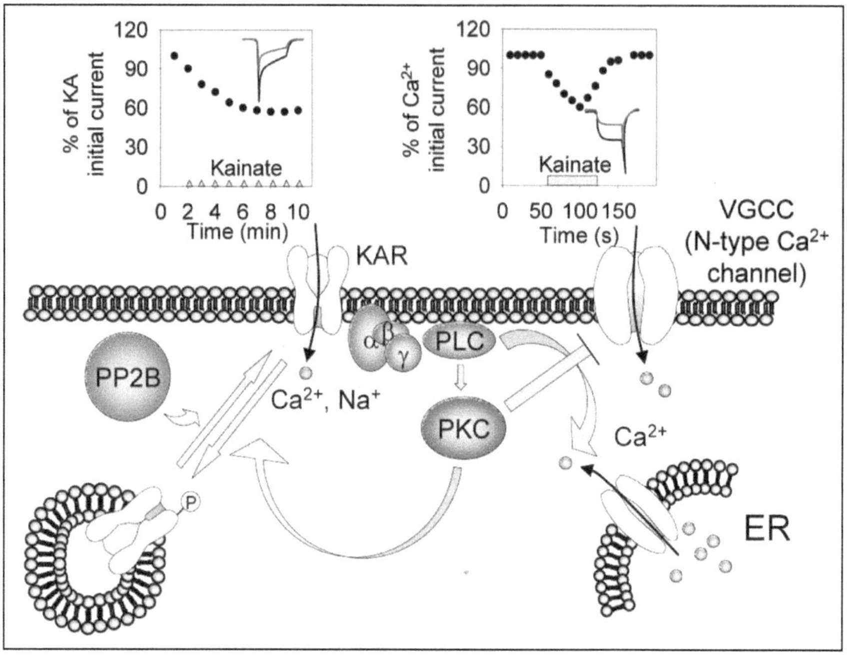

Figure 2. Kainate receptor activates a metabotropic signaling cascade. Kainate receptor induces PKC activation through a pertussis toxin sensitive G-protein and PLC. Besides the release of Ca^{2+} from intracellular stores, the metabotropic cascade leads to the VGCC inhibition. PKC activity also induces kainate receptor internalization when it is overactivated. Calcineurin (PP2B) counteracts this pathway through receptor dephosphorylation.

translates into a general decrease of plasma membrane resistance causing a "membrane shunting". However, in some DRG neurons with low, or even absent kainate ionotropic responses and in the absence of extracellular Na^+, the inhibition of Ca^{2+} influx induced by kainate remains unaffected. The independence of both functions has also been demonstrated by the fact that in cultured DRG neurons, kainate receptors segregate to the neurites and produce inhibition of Ca^{2+} influx at sites different from those points where ionotropic responses (measured in terms of $[Ca^{2+}]$) take place. (2) This noncanonical signalling pathway depends on the GluR5 subunit, because it is absent in DRG neurons from GluR5 KO mice but not from GluR6 KO mice. However, this does not mean that GluR5 is the "bona fide" signal transducer. This ability could resides in the KA1 and/or KA2 subunit, but their delivery to the cell surface could be impaired if either GluR5 or GluR6, necessary to assemble functional receptors, were absent.

As mentioned above, we still do not understand many aspects of the metabotropic cascade. One of the most important questions is how a G-protein could be activated by a ionotropic receptor, which lacks the necessary intracellular molecular motifs. And, of course, it is still necessary to elucidate whether the signal transduction depends on any of the other subunits (KA1 and KA2) as it has been described in other systems (ref. 74, but see also ref. 24).

Kainate Receptors Plasticity

As well as being involved in the modulation of the glutamate release in the dorsal horn, the noncanonical signaling pathway activated by kainate receptors in DRG neurons seems to play a role in its own regulation.

In a similar way as observed in hippocampal cultured neurons where GluR6 subunits undergo a process of insertion into and retrieval from the plasma membrane,[75] in DRG neurons there is a similar system of this regulation. In cultured DRG neurons and heterologous systems, such as SHSY-5Y neuroblastoma cells, it has been shown that the kainate receptor activation of this metabotropic cascade is able to produce the downregulation of the kainate-induced currents.[40]

The decrease of the kainate induced currents is not produced by reducing the receptor conductance, but by the retrieval of receptors from the plasma membrane. Stronger stimulations (repetitive exposure to the agonist with short time for recovery), lead to a 50% decrease in kainate-induced currents. The activation of the receptor entails the activation of a PKC, capable of receptor phosphorylation on its C-terminal, where there are two targets: S879 and S885. This phosphorylation accelerates, ultimately, the receptor internalization. On the other hand, the activation of the phosphatase 2B (PP2B, calcineurin) that presumably dephosphorylates the receptor, prevents current loss. Intracellular [Ca^{2+}] seems to have a key role in this balance of insertion/retrieval. High intracellular [Ca^{2+}] concentrations linked to an activation of the receptor would promote the internalization, whereas low [Ca^{2+}] concentrations would cause an increase of receptors attached to the cell surface. This bidirectional effect could be caused by different affinities for Ca^{2+} exhibited by PKC and by calcineurin; PKC needs higher intracellular [Ca^{2+}] to translocate itself to the membrane, but calcineurin could be activated at much lower [Ca^{2+}][76-78] (see Fig. 2).

In the phenomenon of autoregulation there is special relevance of the alternative splice variants of the GluR5 subunit. GluR5a, the shortest version, lacks the C-terminal segment where the target serines by PKC are located. On the contrary, GluR5b and GluR5c variants posses the insert of 49 aa where the serines are located. Target sequences are located close to the GluR5 retention signal (LTCHQRRTQ) which regulates the retention of GluR5 in the ER. This sequence is also absent in GluR5a variant.[39]

What is the biological significance of this regulatory mechanism? Kainate receptors are able to sense their own overactivation, so high concentrations of glutamate can cause their auto-internalization and thereby attenuation of the response. This situation has already been described for AMPA receptors and for kainate receptors in other systems (fundamentally in central neurons), such as the hippocampus (during LTP and LTD),[79,80] but in this case, the particular relevance of receptor self-regulation is that it helps to maintain the balance of the system.

Kainate Receptors and Pain

Kainate receptors are present at several levels in nociceptive neurotransmission (see reviews in refs. 81 and 82). They are expressed in DRG neurons and in the dorsal horn neurons of the spinal cord. They are also present at the inhibitory synaptic terminals of dorsal horn inhibitory interneurons. Therefore, although kainate receptors influence in pain transmission has been demonstrated, some surprising findings have raised doubts about their specific role in nociception.

Inside DRGs, kainate receptors are located specifically in the small size neurons, identified as the somata of the C fibers (nonmyelinated fibers). They are positive cells for IB4 (isolectin B4) marker and negatives for substance P.[46] The sensory information carried by these fibers is nociceptive and thermoceptive: 60% of these neurons also show expression of VR1 receptors, mediators of noxious heat sensation.[46] One of the main difficulties in the analysis of kainate receptors function in pain is the differentiation between postsynaptic effects (mediated by GluR6 and/or heteromers of this subunit) and presynaptic effects (fundamentally mediated by GluR5 or GluR5/KA2),[26] specially when many tests should be performed in vivo.

The results obtained by using agonists and antagonist of kainate receptors are, in some cases, surprising. In general, kainate receptors antagonists show antinociceptive actions, for example, with SYM 2081 in models of hot plate and mechanical pain.[83] However, SYM 2081, which is a weak antagonist of kainate receptors including GluR6 subunits, is probably inducing a mix pre and postsynaptic blockade. On the other hand, LY382884, the specific antagonist of GluR5, does not cause any change in the hot-plate test,[84] but it has antinociceptive effects in the formalin test (used as model of inflammatory pain).[85] Surprisingly, ATPA, an agonist of GluR5 containing receptors, is also able to reduce nociceptive reflexes in vitro.[84]

Although the tests in knockout mice are still insufficient, they have partially confirmed the results obtained using antagonists. The response to formalin injections and capsaicin is almost abolished in GluR5, but not in GluR6 knock-out mice,[86] showing that sensory transmission of that stimulus is mediated via DRG neurons and with the participation of the presynaptic kainate receptors.

As already mentioned, in genetically modified mice with GluR5 subunit permanently edited in R form, mice do not show any alteration in the sensorial painful transmission, either in hot plate test or in the formalin test,[47] despite a remarkable decrease of the kainate-induced current. This could indicate that it is the presence of the receptor and not its ionotropic activity or Ca^{2+} influx through the channel, that is the necessary condition to support the sensory transmission. Another possibility to consider is the substitution of the subunit by another one, as occurs in hippocampal slices,[24] but not in DRGs cultures.

In summary, antagonists of kainate receptors-containing GluR5 subunit have antinociceptive potential. However an excessive stimulation of the receptor could induce its own inhibition, presumably due receptor internalization. A more extensive study of some of these surprising aspects is necessary. For example, the importance of the editing process or the implication of the noncanonical signalling pathway in pain transmission remains to be determined.

Conclusion

Important progress has been achieved over the last decade on the knowledge of the biology of the kainate receptors and also on their role in synaptic transmission at DRG neurons. However, there are still many intriguing questions to answer. One of those is related to how it is possible that an ionotropic receptor can activates a metabotropic cascade and which are the proteins that mediate this interaction (reviewed in refs. 87-89). This is probably the most attractive aspect and the one which has raised the biggest controversy, mainly in the CNS, but also at peripheral neurons. Nevertheless, concrete steps have already been taken to answer this question: high affinity kainate subunits such as KA2 may be responsible of this interaction.[74] We do not know much about the other components of the signaling pathway either (which isoform of PKC is involved downstream the receptor activation?) or about the function of Ca^{2+} released from the intracellular stores.

The study of targeting and plasticity of synaptic receptors has aroused an increasing interest, being currently a very active field of research and this aspect of kainate receptors physiology is not an exception.[40,90] Understanding how the number of receptors is regulated at the plasma membrane will allow us to know their effective influence in the synaptic transmission and in this particular case, in the transmission of sensorial painful information. In this regard, to elucidate the machinery (and the components of this machinery) that is involved in receptor internalization and delivery would be of critical importance.

Finally, it is necessary to highlight that, although kainate receptors has been demonstrated to be implicated in pain transmission and antagonist of this receptors has mostly antinociceptive effects, further investigations are necessary. It has been suggested that their modulatory actions are exerted for long lasting pain,[81,86] but mechanisms such as compensation by other receptors in KO mice need to be investigated. From another point of view, kainate receptors could be lost, in adulthood, along with some functions related to their function during development or maturation,[91] such as during axon guidance.[92] These functions would be progressively reduced during the establishment of mature synapses. However, these hypotheses will need confirmations over the coming years.

Acknowledgements

I am very grateful to Dr. Rafael Fernández Chacón for his suggestions and critical reading of the manuscript. Also I am indebted to Dr. Juan Lerma, for his comments and of course, for his guidance through the study of kainate receptors biology.

References

1. Gu JG, MacDermott AB. Activation of ATP P2X receptors elicits glutamate release from sensory neuron synapses. Nature 1997; 389:749-753.
2. Ahluwalia J, Urban L, Capogna M et al. Cannabinoid 1 receptors are expressed in nociceptive primary sensory neurons. Neuroscience 2000; 100:685-688.
3. Michael GJ, Priestley JV. Differential expression of the mRNA for the vanilloid receptor subtype 1 in cells of the adult rat dorsal root and nodose ganglia and its downregulation by axotomy. J Neurosci 1999; 19:1844-1854.
4. White G. GABAA-receptor-activated current in dorsal root ganglion neurons freshly isolated from adult rats. J Neurophysiol 1990; 64:57-63.
5. Agrawal SG, Evans RH. The primary afferent depolarizing action of kainate in the rat. Br J Pharmacol 1986; 87:345-355.
6. Lucifora S, Willcockson HH, Lu CR et al. Presynaptic low- and high-affinity kainate receptors in nociceptive spinal afferents. Pain 2006; 120:97-105.
7. Huettner JE. Glutamate receptor channels in rat DRG neurons: activation by kainate and quisqualate and blockade of desensitization by Con A. Neuron 1990; 5:255-266.
8. Bettler B, Boulter J, Hermans-Borgmeyer I et al. Cloning of a novel glutamate receptor subunit, GluR5: expression in the nervous system during development. Neuron 1990; 5:583-595.
9. Bettler B, Egebjerg J, Sharma G et al. Cloning of a putative glutamate receptor: a low affinity kainate-binding subunit. Neuron 1992; 8:257-265.
10. Sommer B, Burnashev N, Verdoorn TA et al. A glutamate receptor channel with high affinity for domoate and kainate. Embo J 1992; 11:1651-656.
11. Gregor P, O'Hara BF, Yang X et al. Expression and novel subunit isoforms of glutamate receptor genes GluR5 and GluR6. Neuroreport 1993; 4:1343-1346.
12. Werner P, Voigt M, Keinanen K et al. Cloning of a putative high-affinity kainate receptor expressed predominantly in hippocampal CA3 cells. Nature 1991; 351:742-744.
13. Paternain AV, Morales M, Lerma J. Selective antagonism of AMPA receptors unmasks kainate receptor-mediated responses in hippocampal neurons. Neuron 1995; 14:185-189.
14. Wilding TJ, Huettner JE. Differential antagonism of alpha-amino-3-hydroxy-5-methyl-4-isoxazolepropionic acid-preferring and kainate-preferring receptors by 2,3-benzodiazepines. Mol Pharmacol 1995; 47:582-587.
15. Smolders I, Bortolotto ZA, Clarke VR et al. Antagonists of GLU(K5)-containing kainate receptors prevent pilocarpine-induced limbic seizures. Nat Neurosci 2002; 5:796-804.
16. Bortolotto ZA, Clarke VR, Delany CM et al. Kainate receptors are involved in synaptic plasticity. Nature 1999; 402:297-301.
17. Bureau I, Bischoff S, Heinemann SF et al. Kainate receptor-mediated responses in the CA1 field of wild-type and GluR6-deficient mice. J Neurosci 1999; 19:653-663.
18. Contractor A, Swanson GT, Sailer A et al. Identification of the kainate receptor subunits underlying modulation of excitatory synaptic transmission in the CA3 region of the hippocampus. J Neurosci 2000; 20:8269-8278.
19. Contractor A, Sailer AW, Darstein M et al. Loss of kainate receptor-mediated heterosynaptic facilitation of mossy-fiber synapses in KA2−/− mice. J Neurosci 2003; 23:422-429.
20. Darstein M, Petralia RS, Swanson GT et al. Distribution of kainate receptor subunits at hippocampal mossy fiber synapses. J Neurosci 2003; 23:8013-8019.
21. Wisden W, Seeburg PH. A complex mosaic of high-affinity kainate receptors in rat brain. J Neurosci 1993; 13:3582-3598.
22. Mulle C, Sailer A, Swanson GT et al. Subunit composition of kainate receptors in hippocampal interneurons. Neuron 2000; 28:475-484.
23. Paternain AV, Herrera MT, Nieto MA et al. GluR5 and GluR6 kainate receptor subunits coexist in hippocampal neurons and coassemble to form functional receptors. J Neurosci 2000; 20:196-205.
24. Christensen JK, Paternain AV, Selak S et al. A mosaic of functional kainate receptors in hippocampal interneurons. J Neurosci 2004; 24:8986-8993.
25. Partin KM, Patneau DK, Winters CA et al. Selective modulation of desensitization at AMPA versus kainate receptors by cyclothiazide and concanavalin A. Neuron 1993; 11:1069-1082.
26. Kerchner GA, Wilding TJ, Huettner JE et al. Kainate receptor subunits underlying presynaptic regulation of transmitter release in the dorsal horn. J Neurosci 2002; 22:8010-8017.
27. Lee CJ, Bardoni R, Tong CK et al. Functional expression of AMPA receptors on central terminals of rat dorsal root ganglion neurons and presynaptic inhibition of glutamate release. Neuron 2002; 35:135-146.
28. Lee CJ, Labrakakis C, Joseph DJ et al. Functional similarities and differences of AMPA and kainate receptors expressed by cultured rat sensory neurons. Neuroscience 2004; 129:35-48.

29. Herb A, Burnashev N, Werner P et al. The KA-2 subunit of excitatory amino acid receptors shows widespread expression in brain and forms ion channels with distantly related subunits. Neuron 1992; 8:775-785.
30. Sahara Y, Noro N, Iida Y et al. Glutamate receptor subunits GluR5 and KA-2 are coexpressed in rat trigeminal ganglion neurons. J Neurosci 1997; 17:6611-6620.
31. Lerma J, Paternain AV, Rodriguez-Moreno A et al. Molecular physiology of kainate receptors. Physiol Rev 2001; 81:971-998.
32. Pinheiro P, Mulle C. Kainate receptors. Cell Tissue Res 2006; 326:457-482.
33. Lerma J. Kainate receptor physiology. Curr Opin Pharmacol 2006; 6:89-97.
34. Thomas NK, Hawkins LM, Miller JC et al. Pharmacological differentiation of kainate receptors on neonatal rat spinal motoneurones and dorsal roots. Neuropharmacology 1998; 37:1223-1237.
35. Laezza F, Wilding TJ, Sequeira S et al. KRIP6: a novel BTB/kelch protein regulating function of kainate receptors. Mol Cell Neurosci 2007; 34:539-550.
36. Garcia EP, Mehta S, Blair LA et al. SAP90 binds and clusters kainate receptors causing incomplete desensitization. Neuron 1998; 21:727-739.
37. Huntley GW, Rogers SW, Moran T et al. Selective distribution of kainate receptor subunit immunoreactivity in monkey neocortex revealed by a monoclonal antibody that recognizes glutamate receptor subunits GluR5/6/7. J Neurosci 1993; 13:2965-2981.
38. Hwang SJ, Pagliardini S, Rustioni A et al. Presynaptic kainate receptors in primary afferents to the superficial laminae of the rat spinal cord. J Comp Neurol 2001; 436:275-289.
39. Ren Z, Riley NJ, Needleman LA et al. Cell surface expression of GluR5 kainate receptors is regulated by an endoplasmic reticulum retention signal. J Biol Chem 2003; 278:52700-52709.
40. Rivera R, Rozas JL, Lerma J. PKC-dependent autoregulation of membrane kainate receptors. Embo J 2007; 26:4359-4367.
41. Higuchi M, Single FN, Kohler M et al. RNA editing of AMPA receptor subunit GluR-B: a base-paired intron-exon structure determines position and efficiency. Cell 1993; 75:1361-1370.
42. Herb A, Higuchi M, Sprengel R et al. Q/R site editing in kainate receptor GluR5 and GluR6 premRNAs requires distant intronic sequences. Proc Natl Acad Sci USA 1996; 93:1875-1880.
43. Burnashev N, Zhou Z, Neher E et al. Fractional calcium currents through recombinant GluR channels of the NMDA, AMPA and kainate receptor subtypes. J Physiol 1995; 485(Pt 2):403-418.
44. Swanson GT, Feldmeyer D, Kaneda M et al. Effect of RNA editing and subunit coassembly single-channel properties of recombinant kainate receptors. J Physiol 1996; 492(Pt 1):129-142.
45. Bernard A, Ferhat L, Dessi F et al. Q/R editing of the rat GluR5 and GluR6 kainate receptors in vivo and in vitro: evidence for independent developmental, pathological and cellular regulation. Eur J Neurosci 1999; 11:604-616.
46. Lee CJ, Kong H, Manzini MC et al. Kainate receptors expressed by a subpopulation of developing nociceptors rapidly switch from high to low Ca^{2+} permeability. J Neurosci 2001; 21:4572-4581.
47. Sailer A, Swanson GT, Perez-Otano I et al. Generation and analysis of GluR5(Q636R) kainate receptor mutant mice. J Neurosci 1999; 19:8757-8764.
48. Kerchner GA, Wilding TJ, Li P et al. Presynaptic kainate receptors regulate spinal sensory transmission. J Neurosci 2001; 21:59-66.
49. Li P, Wilding TJ, Kim SJ et al. Kainate-receptor-mediated sensory synaptic transmission in mammalian spinal cord. Nature 1999; 397:161-164.
50. Kerchner GA, Wang GD, Qiu CS et al. Direct presynaptic regulation of GABA/glycine release by kainate receptors in the dorsal horn: an ionotropic mechanism. Neuron 2001; 32:477-488.
51. Xu H, Wu LJ, Zhao MG et al. Presynaptic regulation of the inhibitory transmission by GluR5-containing kainate receptors in spinal substantia gelatinosa. Mol Pain 2006; 2:29.
52. Rodriguez-Moreno A, Herreras O, Lerma J. Kainate receptors presynaptically downregulate GABAergic inhibition in the rat hippocampus. Neuron 1997; 19:893-901.
53. Rodriguez-Moreno A, Lerma J. Kainate receptor modulation of GABA release involves a metabotropic function. Neuron 1998; 20:1211-1218.
54. Rozas JL, Paternain AV, Lerma J. Noncanonical signaling by ionotropic kainate receptors. Neuron 2003; 39:543-553.
55. Cunha RA, Malva JO, Ribeiro JA. Pertussis toxin prevents presynaptic inhibition by kainate receptors of rat hippocampal ((3)H)GABA release. FEBS Lett 2000; 469:159-162.
56. Frerking M, Schmitz D, Zhou Q et al. Kainate receptors depress excitatory synaptic transmission at CA3—>CA1 synapses in the hippocampus via a direct presynaptic action. J Neurosci 2001; 21:2958-2966.
57. Melyan Z, Wheal HV, Lancaster B. Metabotropic-mediated kainate receptor regulation of IsAHP and excitability in pyramidal cells. Neuron 2002; 34:107-114.

58. Streb H, Irvine RF, Berridge MJ et al. Release of Ca^{2+} from a nonmitochondrial intracellular store in pancreatic acinar cells by inositol-1,4,5-trisphosphate. Nature 1983; 306:67-69.
59. Burgess GM, Godfrey PP, McKinney JS et al. The second messenger linking receptor activation to internal Ca release in liver. Nature 1984; 309:63-66.
60. Berridge MJ, Bootman MD, Roderick HL. Calcium signalling: dynamics, homeostasis and remodelling. Nat Rev Mol Cell Biol 2003; 4:517-529.
61. Clapham DE. Calcium signaling. Cell 2007; 131:1047-1058.
62. Garcia AG, Garcia-De-Diego AM, Gandia L et al. Calcium signaling and exocytosis in adrenal chromaffin cells. Physiol Rev 2006; 86:1093-1131.
63. Jahn R, Lang T, Sudhof TC. Membrane fusion. Cell 2003; 112:519-533.
64. Lauri SE, Bortolotto ZA, Nistico R et al. A role for Ca^{2+} stores in kainate receptor-dependent synaptic facilitation and LTP at mossy fiber synapses in the hippocampus. Neuron 2003; 39:327-341.
65. Abeliovich A, Paylor R, Chen C et al. PKC gamma mutant mice exhibit mild deficits in spatial and contextual learning. Cell 1993; 75:1263-1271.
66. Rhee JS, Betz A, Pyott S et al. Beta phorbol ester- and diacylglycerol-induced augmentation of transmitter release is mediated by Munc13s and not by PKCs. Cell 2002; 108:121-133.
67. Brose N, Rosenmund C. Move over protein kinase C, you've got company: alternative cellular effectors of diacylglycerol and phorbol esters. J Cell Sci 2002; 115:4399-4411.
68. Gross RA, MacDonald RL. Activators of protein kinase C selectively enhance inactivation of a calcium current component of cultured sensory neurons in a pertussis toxin-sensitive manner. J Neurophysiol 1989; 61:1259-1269.
69. Zong X, Lux HD. Augmentation of calcium channel currents in response to G protein activation by GTP gamma S in chick sensory neurons. J Neurosci 1994; 14:4847-4853.
70. Van Kolen K, Pullan S, Neefs JM et al. Nociceptive and behavioural sensitisation by protein kinase Cepsilon signalling in the CNS. J Neurochem 2008; 104:1-13.
71. Jarvis SE, Zamponi GW. Distinct molecular determinants govern syntaxin 1A-mediated inactivation and G-protein inhibition of N-type calcium channels. J Neurosci 2001; 21:2939-2948.
72. Herlitze S, Zhong H, Scheuer T et al. Allosteric modulation of Ca^{2+} channels by G proteins, voltage-dependent facilitation, protein kinase C and Ca(v)beta subunits. Proc Natl Acad Sci USA 2001; 98:4699-4704.
73. Peterson BZ, DeMaria CD, Adelman JP et al. Calmodulin is the Ca^{2+} sensor for Ca^{2+} -dependent inactivation of L-type calcium channels. Neuron 1999; 22:549-558.
74. Ruiz A, Sachidhanandam S, Utvik JK et al. Distinct subunits in heteromeric kainate receptors mediate ionotropic and metabotropic function at hippocampal mossy fiber synapses. J Neurosci 2005; 25:11710-11718.
75. Martin S, Henley JM. Activity-dependent endocytic sorting of kainate receptors to recycling or degradation pathways. Embo J 2004; 23:4749-4759.
76. Stewart AA, Ingebritsen TS, Manalan A et al. Discovery of a Ca^{2+}- and calmodulin-dependent protein phosphatase: probable identity with calcineurin (CaM-BP80). FEBS Lett 1982; 137:80-84.
77. Perrino BA, Ng LY, Soderling TR. Calcium regulation of calcineurin phosphatase activity by its B subunit and calmodulin. Role of the autoinhibitory domain. J Biol Chem 1995; 270:7012.
78. Kohout SC, Corbalan-Garcia S, Torrecillas A et al. C2 domains of protein kinase C isoforms alpha, beta and gamma: activation parameters and calcium stoichiometries of the membrane-bound state. Biochemistry 2002; 41:11411-11424.
79. Takahashi T, Svoboda K, Malinow R. Experience strengthening transmission by driving AMPA receptors into synapses. Science 2003; 299:1585-1588.
80. Lee SH, Liu L, Wang YT et al. Clathrin adaptor AP2 and NSF interact with overlapping sites of GluR2 and play distinct roles in AMPA receptor trafficking and hippocampal LTD. Neuron 2002; 36:661-674.
81. Ruscheweyh R, Sandkuhler J. Role of kainate receptors in nociception. Brain Res Brain Res Rev 2002; 40:215-222.
82. Wu LJ, Ko SW, Zhuo M. Kainate receptors and pain: from dorsal root ganglion to the anterior cingulate cortex. Curr Pharm Des 2007; 13:1597-1605.
83. Sutton JL, Maccecchini ML, Kajander KC. The kainate receptor antagonist 2S,4R-4-methylglutamate attenuates mechanical allodynia and thermal hyperalgesia in a rat model of nerve injury. Neuroscience 1999; 91:283-292.
84. Procter MJ, Houghton AK, Faber ES et al. Actions of kainate and AMPA selective glutamate receptor ligands on nociceptive processing in the spinal cord. Neuropharmacology 1998; 37:1287-1297.
85. Simmons RM, Li DL, Hoo KH et al. Kainate GluR5 receptor subtype mediates the nociceptive response to formalin in the rat. Neuropharmacology 1998; 37:25-36.
86. Ko S, Zhao MG, Toyoda H et al. Altered behavioral responses to noxious stimuli and fear in glutamate receptor 5 (GluR5)- or GluR6-deficient mice. J Neurosci 2005; 25:977-84.

87. Lerma J. Roles and rules of kainate receptors in synaptic transmission. Nat Rev Neurosci 2003; 4:481-95.
88. Rodriguez-Moreno A, Sihra TS. Kainate receptors with a metabotropic modus operandi. Trends Neurosci 2007; 30:630-67.
89. Rodriguez-Moreno A, Sihra TS. Metabotropic actions of kainate receptors in the CNS. J Neurochem 2007; 103:2121-2135.
90. Martin S, Nishimune A, Mellor JR et al. SUMOylation regulates kainate-receptor-mediated synaptic transmission. Nature 2007; 447:321-325.
91. Marchal C, Mulle C. Postnatal maturation of mossy fibre excitatory transmission in mouse CA3 pyramidal cells: a potential role for kainate receptors. J Physiol 2004; 561:27-37.
92. Ibarretxe G, Perrais D, Jaskolski F et al. Fast regulation of axonal growth cone motility by electrical activity. J Neurosci 2007; 27:7684-7695.

CHAPTER 8

Role of Kainate Receptors in Network Activity during Development

Sari E. Lauri* and Tomi Taira

Abstract

Distinct populations of kainate-type ionotropic glutamate receptors (KARs), located at various cell types and subcellular compartments and utilizing diverse downstream signaling mechanisms, represent an intricate system with large capacity for modulatory effects ranging from synapse-specific changes to alterations in the excitability of large neuronal ensembles. However, the way the diverse functions ascribed for KARs are utilized under different physiological and pathological conditions to regulate activity at the level of neuronal networks is still largely unclear. Here, we address the data regarding functions of KARs in the regulation of network activity in the hippocampus, with a main focus on their roles during early postnatal development. We further discuss the evidence suggesting that KAR mediated signaling during the immature type network activity is involved in the formation and maturation of glutamatergic synapses.

Introduction

During early development, immature neuronal networks typically display spontaneous, rhythmic activity which is characterized by short (tens to hundreds of milliseconds) bursts of synchronous activity occurring at intervals of tens of seconds (e.g., hippocampus) to few minutes (e.g., retina, spinal cord).[1,2] During the bursts, ensembles of neurons fire together at a high frequency, thus enabling temporally and spatially correlated electrical signaling in the immature circuitry. Accumulating evidence suggests that the immature network activity is instrumental for the development of the synaptic connectivity.[2,3] Electrical activity is not essential for synapse formation per se; both the pre and postsynaptic specializations form constitutively in the absence of activity even in vivo (e.g., ref. 4). However, it is evident that activity modulates stabilization and elimination of synapses and neuronal branches to guide refinement and remodeling of the neuronal connectivity.[5-7]

The mechanisms generating rhythmic activity in the immature brain differ depending on the brain area and on the developmental stage. In the rat hippocampus, the early synchronous activity around the time of birth is mainly dependent on gap junctions.[8] These early oscillations are rapidly replaced by synaptically driven network bursts, which critically depend on both GABAergic and glutamatergic transmission[9-11] and govern the activity in the immature hippocampus until postnatal day 10-12.[12]

Recent findings from several groups have indicated that kainate-type of glutamate receptors (KARs) have concentration, cell type and cell compartment specific effects on the function of both glutamatergic and GABAergic neurons (for reviews see refs. 13-15). Thus, postsynaptic KARs

*Corresponding Author: Sari E. Lauri—Neuroscience Center and Department of Bio- and Environmental Sciences, P.O. Box 65 (Viikinkaari 1), 00014 University of Helsinki, Finland. Email: sari.lauri@helsinki.fi

Kainate Receptors: Novel Signaling Insights, edited by Antonio Rodríguez-Moreno and Talvinder S. Sihra. ©2011 Landes Bioscience and Springer Science+Business Media.

typically mediate a slow, small amplitude postsynaptic current (EPSC), while presynaptic KARs act as auto- or heteroreceptors regulating transmitter release (e.g., references 16-20). Synaptic activation of KARs typically shows remarkable dependency on stimulation frequency, which together with the kinetic properties of the receptor provides means for integrating and transmitting information of the afferent firing rate at the central synapses.[13-15] Apart from its synaptic functions, KAR activation has profound effects on neuronal excitability by regulating the afterhyperpolarizing potassium currents in certain neurons.[21,22] These effects combined to the modulatory functions at both GABA- and glutamatergic synapses enable KARs to efficiently regulate transmission and excitability at the level of neuronal networks.[19,23,24]

Kainate Receptors and Network Activity in the Mature Hippocampus

Kainate (KA) application is potent way to induce maintained pathological synchrony and has been widely studied as animal model for epileptogenesis in vivo.[25,26] Unfortunately, only a minority of the published data concerning the mechanisms of KA induced network activity address selectively the role of KARs, due to overlap in the pharmacological properties of AMPA-Rs and KARs. Thus, kainate at relatively low concentrations activates not only KARs but also produces a nondesensitizing response at AMPA receptors,[27-30] which may efficiently increase excitability in the recurrent circuits in area CA3.

Strong evidence for contribution of KARs in the generation of pathological synchrony in vivo originates from GluR6-deficient mice, which are less susceptible to epileptogenesis in response to systemic administration of kainate than the wild-type animals.[31] Further, GluR5 antagonism has been shown to prevent pilocarpine-induced limbic seizures[32] while infusion of GluR5 agonist induces limbic seizures, probably due to its actions in the amygdala.[33,34] Recently, GluR5 antagonist was shown to reduce the frequency of hippocampal theta-oscillations (5-12 Hz) in freely moving animals.[35]

Results supporting involvement of KARs in hippocampal network oscillations have also been obtained in vitro. Kainate application in the hippocampal slices generates rhythmic oscillations in the gamma frequency range (30-100 Hz), similar to those observed in the brain under physiological conditions.[36] These oscillations are strongly reduced or absent in area CA3 of hippocampal slices from GluR6 –/– mice, while genetic ablation of GluR5 leads to a higher susceptibility of the network to the oscillogenic effects of kainate.[23] In slices from GluR7–/– mice, kainate-induced gamma oscillations were indistinguishable from wild-type.[23] These phenotypes suggest opposite roles for the KAR subunits GluR5 and GluR6 in kainate induced gamma-oscillations. However, other studies have shown that GluR5 antagonists reduce the power of pre-established KA induced gamma oscillations in rat hippocampal slices, although activation of these receptors is not sufficient for generating this type of rhythmic activity.[37] One explanation for the discrepancy is functional compensation and alterations in subcellular trafficking of kainate receptor subunits in the knockout mice.[38] In particular, heteromeric GluR5/GluR6 receptors may be blocked by GluR5 antagonists but also functionally deficient in the GluR6–/– mice, due to impaired surface expression of GluR5 alone.[39] Thus, GluR5 containing receptors, possibly expressed as heteromeric combinations with GluR6, may play a role in maintaining kainate-driven gamma frequency oscillations and in regulation of the oscillation power.[37] Activation of the GluR6 subunit containing KARs, on the other hand, appears critical for generation of the oscillatory activity.[23]

Synaptic and Cellular Mechanisms

Although the existing data strongly supports a role for KARs in regulation of pathological and physiological network activity in the hippocampus, the synaptic mechanisms underlying these effects are less clear. In general, rhythmic activity and synchronization in neuronal networks can be generated by recurrent excitatory connections, which are found between pyramidal neurons in the area CA3. However, it is well established that pacing of physiological cortical rhythms critically depends upon inhibition originating from interneurons to balance excitation and control the precision of spike timing.[40-42]

Most of the existing data on the mechanisms by which KAR influence gamma oscillations are focused on regulation of interneuronal function. Activation of somatodendritic KARs depolarizes interneurons in both area CA1 and CA3 and leads to strong increase in spontaneous GABAergic activity (sIPSCs).[23,38,43-46] In addition, KARs increase excitability of interneuron axons in area CA1, which might contribute to regulation of sIPSCs.[47] The lack of KA induced neuronal depolarization and regulation of sIPSCs in area CA3 of GluR6–/– mice was proposed to explain the impaired gamma frequency oscillations.[23] Unfortunately, pharmacological data on the role of GluR5 in kainate-induced sIPSCs in area CA3 is unavailable, leaving it unclear whether the observed effects of GluR5 antagonists on the maintenance of gamma oscillations[37] are associated with reduced spontaneous GABAergic activity.

Modulation of the oscillatory activity in the theta frequency range by GluR5 containing KARs was proposed to be due to reduced excitatory input to one or more classes of interneurons in the hippocampus.[35] In line with this idea, synaptic kainate currents in oriens-lacunosum moleculare (O-LM) interneurons in area CA1 may reset their firing phase and have been implicated in generation of theta activity in vitro.[48,49] Further, GluR5 is highly expressed in these neurons.[50]

In addition to its effects on interneuronal excitability, KAR activation strongly modulates glutamatergic transmission at the mossy fibres.[13-15] High frequency activity of the mossy fibres effectively discharges both the pyramidal neurons and interneurons in area CA3 and thereby profoundly influences the activity of the CA3 network.[51] Kainate-receptors promote frequency-dependent facilitation of mossy fibre input to pyramidal neurons during high-frequency transmission,[18,20] which is an important mechanisms to overcome the strong feed-forward inhibition in the circuitry.[52] In parallel, activation of postsynaptic KARs enhance the excitability of pyramidal neurons by mediating slow EPSCs[16,17] and by suppressing the afterhyperpolarizing potassium current (I_{AHP}).[21,22] Together, these mechanisms promote the excitation of CA3 pyramidal neurons in response to high-frequency mossy fiber input, which may critically contribute to the excitation necessary to drive the associative network in area CA3.

Kainate Receptors in the Developing Brain

KARs subunits are highly expressed in the brain during early postnatal development and patterned changes in their cellular and subcellular expression profile take place during maturation.[53,54] In addition, electrophysiological analysis has revealed developmental changes in kainate receptor function in several areas of the brain. For example, at the thalamocortical synapses in the developing rodent barrel cortex, the contribution of KARs to synaptic transmission decreases with a developmental profile that correlates with the critical period for experience-dependent plasticity.[55-56] At the spinal cord, GluR5 containing KARs rapidly switch from high to low calcium permeability, due to RNA editing in the dorsal root ganglion cells, in parallel to synapse formation.[57] In the neonate hippocampus, the tonic activation of presynaptic kainate receptors regulating glutamate release is lost during early postnatal development in parallel with activity -dependent maturation of the circuitry.[19,58]

The developmental switch in KAR function in the hippocampus is associated with a loss of high-affinity receptors[58] as well as a change in their downstream signaling mechanisms.[59] For example, at the immature CA3-CA1 synapses (i.e., postnatal day (P)3-6), activation of presynaptic KARs by a GluR5 selective agonist ATPA leads to depression of both evoked and action potential-independent glutamatergic transmission (mEPSCs) in a G-protein and PKC dependent manner. At P14, the effect of ATPA on mEPSCs is completely lost[59] (Fig. 1), although a strong G-protein dependent but PKC independent depression of evoked glutamatergic transmission is still seen.[60-62] Whether the switch in the signalling mechanisms reflects developmental alterations in the subunit composition, cellular localization or C-terminal coupling of the KARs or independent developmental alterations in the release machinery is not known.

Evidence for developmental downregulation of the presynaptic effects of KARs has also been obtained at GABAergic synapses in the area CA1.[63] In the neonatal CA1, kainate receptor agonists strongly decrease mIPSC frequency,[63] an effect that is larger and much more robust

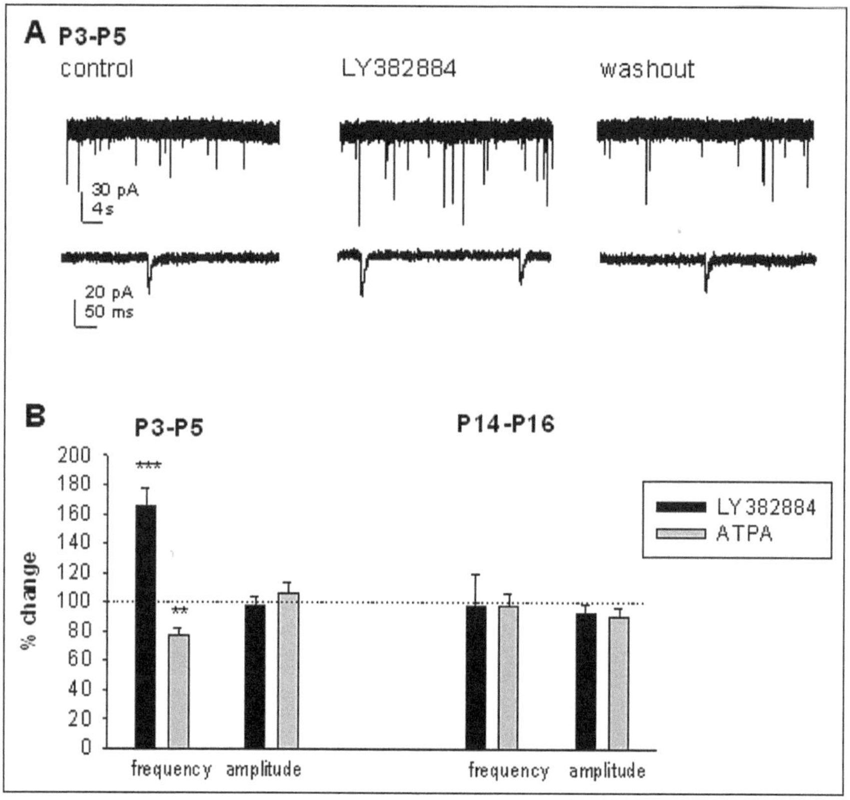

Figure 1. Developmental loss of KAR mediated regulation of mEPSCs in CA1 pyramidal neurons. A) Example traces showing the effect of LY382884 (10 μM) on mEPSCs at CA1 pyramidal neurons at P4. B) Averaged data on the effect of GluR5 selective agonist ATPA (1 μM) and antagonist LY382884 (10 μM) on mEPSC frequency and amplitude at P3-P6 and P14-P16. ** $p < 0.01$; *** $p < 0.005$. Adapted from Lauri SE et al. Neuron 2006; 50:415-429[58] and Sallert M, Malkki H, Segerstråle M et al. Neuropharmacology 2007; 52:1354-1365,[59] ©2006 and ©2007 with permission from Elsevier.

than that reported in older animals.[43,64] In other studies, no effect of KAR agonists on mIPSCs in the hippocampus from two to six week old animals was observed,[44,45,65] implying that similar to glutamatergic synapses,[59] the presynaptic regulation of action-potential independent release at GABAergic synapses in area CA1 might be lost during maturation of the circuitry.

Thus, the mechanisms described for KARs in the mature circuits cannot be directly applied to immature networks, where the expression pattern but also the synaptic and cellular signaling of KARs can be profoundly different as compared to the adult. On the other hand, the observed developmental changes in function raise the intriguing possibility that certain KAR functions might be primarily related to the development and maturation of the synaptic connectivity.

KAR Mediated Regulation of Network Activity in the Immature Hippocampus

In contrast to older animals, synchronous network activity is spontaneously generated in the immature hippocampal slices, which allows investigation of its mechanisms without a need for additional pharmacological interventions. Although initially thought to be mainly GABAergic, accumulating evidence indicates that glutamatergic transmission is critical in generating and regulating the activity and excitability of the immature hippocampal network.[9,10,66,67] Early studies indicate that application of kainate can induce epileptiform activity in the area CA3 already around the time of birth both in vivo and in vitro.[67,68] However, the exact roles of KARs in generation and synchronization of the immature network activity are only beginning to emerge (e.g., reference 19).

Effects of KAR Agonists on Spontaneous Activity in the Immature Hippocampus In Vitro

Kainate application at concentrations (25-50 nM) selective for high-affinity kainate receptors can efficiently increase the frequency of spontaneous network bursts in the immature hippocampus without generation of epileptiform hypersynchrony (Juuri, Lauri and Taira, unpublished data). At higher nanomolar concentrations (250-300 nM), kainate induces epileptiform activity already at P2.[67] Intriquingly, these effects are completely different from what is observed by selectively activating GluR5 subunit containing KARs by the agonist ATPA.[19,69] ATPA application at P3-P6 leads to substantial inhibition of the spontaneous network bursts, associated with a large increase in the frequency of sIPSCs and depolarization of interneurons (Fig 2).[19] The effect of ATPA is thus dominated by a shift in a balance towards asynchronous GABAergic transmission, which efficiently inhibits the synchronous network activity. Interestingly, kainate (250 nM-1 µM) application leads to similar effects on GABAergic transmission (e.g., references 43,63), but is also associated with a large increase in glutamatergic drive producing a robust increase in the network

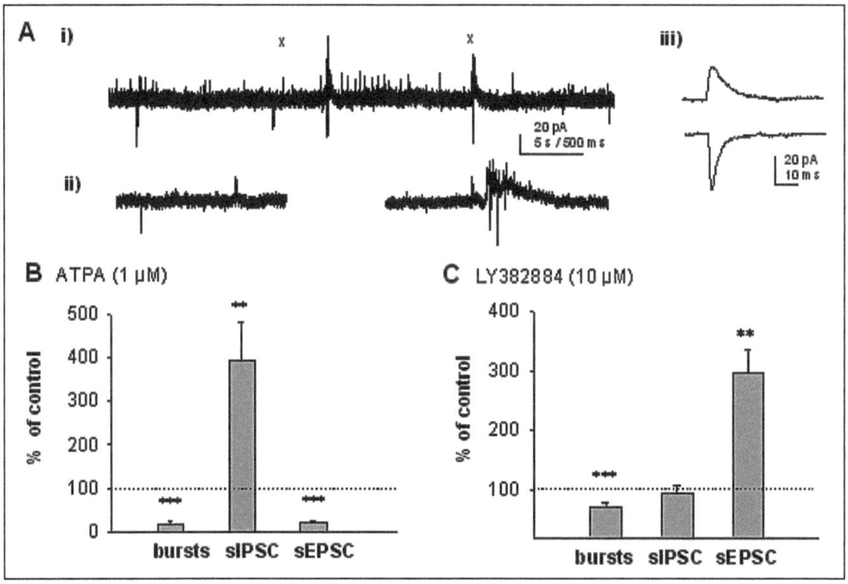

Figure 2. Spontaneous network activity in the immature hippocampus is regulated by GluR5 subunit containing KARs. A) A typical pattern of spontaneous activity recorded from CA3 pyramidal neuron (P4), using a low chloride solution in the patch electrode (i). sIPSCs are seen as outward currents and sEPSCs as inward currents, as is shown in expanded time scale (ii, left and iii). The network bursts consist of a slow GABAergic current and a barrage of EPSCs (ii, right). X marks the place of the expanded time scale trace. Effect of GluR5 subunit selective agonist ATPA (B) and antagonist LY382884 (C) on the frequency of spontaneous bursts, sEPSCs and sIPSCs in CA3 pyramidal neurons (P3-P6). ** $p < 0.01$; *** $p < 0.005$. Adapted from Lauri SE et al. J. Neurosci 2005; 25(18):4473-4484.[19]

excitability.[67] Thus, at immature networks, increase in the GABAergic transmission alone appears not sufficient for burst generation. In contrast, the spontaneous network activity in the immature circuitry evidently depends on the balance between GABAergic and glutamatergic drive as well as on the overall level of synaptic activity. From this perspective, KARs represent an ideal mechanism for modulation of network activity because of its specific effects on GABAergic and glutamatergic synapse populations.

Endogenous Activity of KARs at Immature Networks

Recent evidence suggests that tonically active KARs at immature synapses might play a critical role in generation and maintenance of the spontaneous network activity. Thus, blocking the GluR5 subunit containing KARs by the selective antagonist LY382884 results in a decrease in the occurrence of the network bursts, suggesting that endogenous activation of KARs is involved in the burst initiation or network synchronization (Fig 2).[19]

The endogenous activation of GluR5 subunit containing KARs appears to preferentially regulate transmission at glutamatergic synapses, since the effect of LY382884 was associated with changes in spontaneous glutamatergic transmission (sEPSCs) at both pyramidal neurons and interneurons in the neonate CA3, while no significant effects on transmission at GABAergic synapses were observed.[19] Interestingly, GluR5 subunit containing KARs regulate transmission differentially depending on the cell type. Thus, glutamatergic input to pyramidal cells were tonically inhibited by endogenously active KARs, while tonic facilitatory effect on input to stratum lucidum interneurons was observed. Moreover, these receptor populations are further distinguished based on the downstream signalling mechanisms, the inhibitory effect on release in pyramidal neurons being G-protein dependent, while the facilitatory effect in interneurons being G-protein independent.[19]

The predominant synaptic mechanism by which the endogenous GluR5 activation alters the early network activity is the enhancement of glutamatergic input to interneurons. This is because glutamatergic drive to pyramidal neurons in area CA3 is much weaker as compared to interneurons during early postnatal development, due to the developmentally earlier synaptogenesis at interneurons.[70] In addition to the synaptic effects, we recently found that endogenously active GluR5 subunit containing receptors tonically enhance the excitability of interneurons in the immature but not two week old mouse CA3 (Segerstrale, Lauri, Mulle and Taira, unpublished data). Together, these actions increase the probability for synchronization of the interneuronal network, which manifest as generation of spontaneous bursts at the immature network.

In conclusion, the present data suggests that at the immature hippocampus, endogenous activity of kainate receptors controls the network by tonically regulating the glutamatergic drive as well as cellular excitability. Given the dramatic effects of KAR activation or inhibition on network activity, it appears that the activation level of KARs in the neonatal hippocampus is finely tuned by ambient glutamate to permit the typical rhythmic activity in the immature circuitry.

Role of KARs in Developmental Maturation of Synaptic Connectivity

The finding that certain functions of KARs are restricted to early postnatal stages and correlate with the developmental pattern of synaptogenesis implies a role for these receptors in maturation of the connectivity. Furthermore, the switch from immature to mature type KAR function appears to be controlled by similar and/or parallel activity-dependent mechanisms that have been proposed to guide fine-tuning of the neuronal connectivity during development. For example, induction of LTP at the developing thalamocortical synapses[55] as well as in the immature CA1 synapses[58] leads to rapid decrease in the physiological activation of KARs.

By acting as direct sensors of glutamatergic activity, kainate receptors would be well suited for coupling patterned electrical activity with morphogenesis with high spatial and temporal resolution. Direct evidence for such mechanism comes from a study by Tashiro et al[71] where it was shown that at the hippocampal mossy fibre synapse, KARs mediate the effects of synaptic stimulation on the motility of axonal filopodia, a process which is thought to be involved in

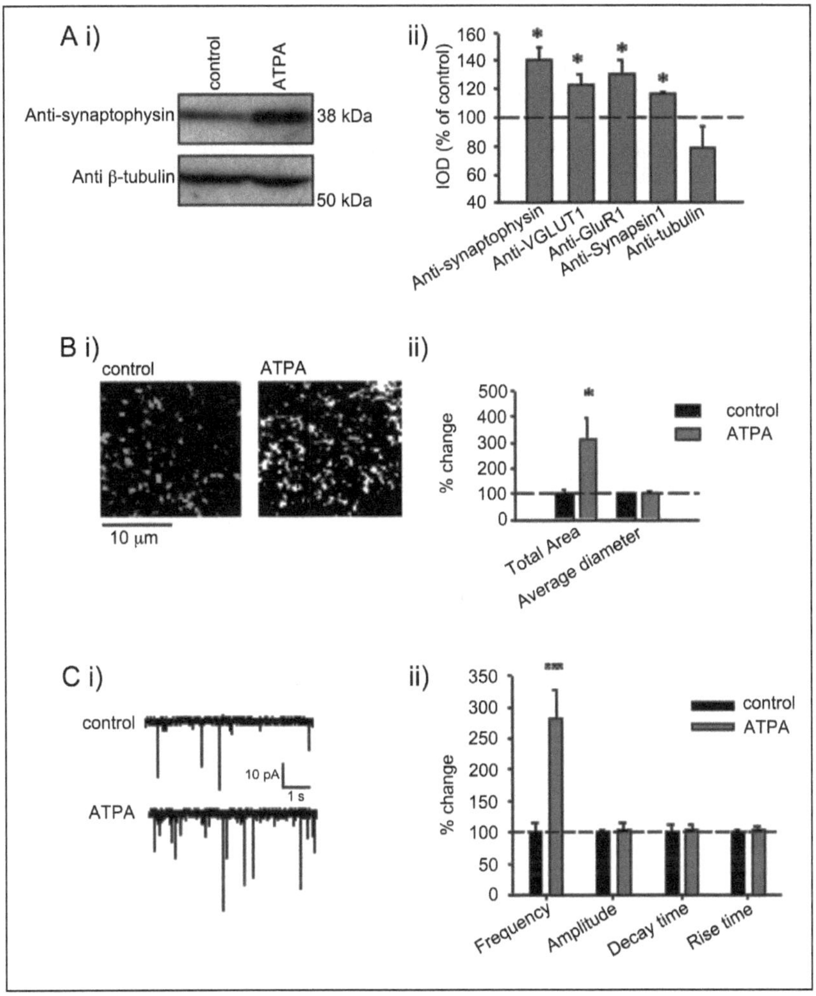

Figure 3. Long-term activation of KARs leads to increase in density of functional synapses in area CA1 of hippocampal slice cultures. A) Western blots (i) and quantified data (ii) showing higher expression level of various synaptic marker proteins in hippocampal slices cultures treated with ATPA (1 μM) for 20 hours as compared to control slices. B) Confocal images (i) and quantified data (ii) of synaptophysin immunofluorescence in the dendritic area of CA1 in ATPA treated vs control slices. *p < 0.05. C) Example traces (i) and analysed data (ii) showing increased frequency but no change in the amplitude or kinetics of AMPA-R-mediated mEPSC in CA1 pyramidal neurons in ATPA treated vs control slices. ***p < 0.005. Adapted with permission from reference 73.

the early steps of synapse formation and rearrangement. At the same synapse, lack of KAR subunits GluR5 and GluR6 during development perturbs establishment of mature of pre and postsynaptic functions.[72]

There is also evidence for a role for the immature-type tonic KAR activity in formation and maturation of glutamatergic circuitry in the area CA1. Mimicking the tonic KAR activity by pharmacological activation of GluR5-containing KARs in slice cultures specifically increases density of functional glutamatergic synapses (Fig. 3).[73] Further, a critical role of endogenous KARs was revealed by long-term treatment of hippocampal cultures with GluR5 antagonist LY382884, which caused a significant impairment of glutamatergic transmission to CA1 pyramidal neurons. The effect of GluR5 activation on synaptic density is dependent on the developmental stage (Vesikansa, Taira and Lauri, unpublished) and is mediated by PKC-coupled signalling of KARs that is only detected at immature CA3-CA1 synapses.[73]

Apart from mediating signals that directly influence the molecular assembly/disassembly of synaptic machinery, KARs may regulate maturation of the circuitry indirectly via its effects on the spontaneous network activity. Inhibition of the synchronous network activity in the immature hippocampus leads to robust increase in the density of functional glutamatergic synapses and parallel decrease in GABAergic synapses.[5,7] These changes in the synaptic connectivity occur in a clearly shorter time scale as compared to homeostatic scaling in mature circuits,[74] thus highlighting the importance of activity in the formation and rearrangement of immature synaptic connectivity.

Conclusion

The recent advances in understanding the functions of KARs demonstrate a critical role for these receptors not only in the epileptiform network activity but also in the physiological oscillations both in the developing and adult hippocampus. Evidently, the effects of KARs at network level are complex and involve multiple mechanisms, reflecting the variability in their synaptic and cellular functions. Further, KAR mediated effects depend on developmental stage and on the prior activity of the network. For example, at the developing hippocampus, physiologically active kainate receptors appear to promote synchronized network activity via regulation of glutamatergic inputs to interneurons and by enhancing interneuron excitability (ref. 19, Segerstrale, Lauri, Mulle Taira, unpublished). These mechanisms, however, are developmentally downregulated and no longer functional at the mature hippocampal network. At older animals, kainate-receptor mediated regulation of gamma frequency oscillations have been mainly attributed to somatodendritic KARs at CA3 interneurons[23,24] while KARs mediating and modulating glutamatergic input at O-LM interneurons have been implicated in the regulation of theta oscillations.[48,35] However, the picture is still far from being complete. For example we know very little on the functions of KARs at different types of interneurons, which might have specific roles in the regulation of network excitability and sychronization under both physiological and pathological conditions.

A key issue in future studies is to understand not only the cellular mechanisms by which KARs modulate synaptic transmission and neuronal function, but also how various KAR populations are activated and regulated under different physiological and pathological conditions. This will require selective genetic or pharmacological manipulation of KARs located in specific neuron types and subcellular compartments in intact neuronal networks, where the physiological balance between excitation and inhibition is not perturbed. Ideally, understanding the network level effects of KARs in detail will provide basis for therapeutic applications to control pathological synchronization and hyperexcitability of neuronal networks.

Acknowledgements

We thank Vernon Clarke for his expert comments on the manuscript. Supported by Academy of Finland and University of Helsinki.

References

1. Feller MB. Spontaneous correlated activity in developing neural circuits. Neuron 1999; 22:653-656.
2. Zhang LI, Poo MM. Electrical activity and development of neural circuits. Nat Neurosci 2001;(4 Suppl) 1207-1214.
3. Ben Ari Y. Excitatory actions of GABA during development: The nature of the nurture. Nat Rev Neurosci 2002; 3:728-739.
4. Verhage M, Maia AS, Plomp JJ et al. Synaptic assembly of the brain in the absence of neurotransmitter secretion. Science 2000; 287:864-869.

5. Lauri SE, Lamsa K, Pavlov I et al. Activity blockade increases the number of functional synapses in the hippocampus of newborn rats. Mol Cell Neurosci 2003; 22:107-117.
6. Hua JY, Smith SJ. Neural activity and the dynamics of central nervous system development. Nat Neurosci 2004; 7(4):327-332.
7. Colin-LeBrun I, Ferrand N, Caillard O et al. Spontaneous synaptic activity is required for formation of functional GABAergic synapses in the developing rat hippocampus. J Physiol 2004; 559:129-139.
8. Crépel V, Aronov D, Jorquera I et al. A parturition-associated nonsynaptic coherent activity pattern in the developing hippocampus. Neuron 2007; 54(1):105-120.
9. Bolea S, Avignone E, Berretta N et al. Glutamate controls the induction of GABA-mediated giant depolarizing potentials through AMPA receptors in neonatal rat hippocampal slices. J Neurophysiol 1999; 81:2095-2102.
10. Lamsa K, Palva JM, Ruusuvuori E et al. Synaptic GABA(A) activation inhibits AMPA-kainate receptor-mediated bursting in the newborn (P0-P2) rat hippocampus. J Neurophysiol 2000; 83:359-366.
11. Palva M, Lamsa K, Lauri SE et al. Fast network oscillations in the newborn rat hippocampus in vitro. J Neurosci 2000; 20(3):1170-1178.
12. Ben-Ari Y, Cherubini E, Corradetti R et al. Giant synaptic potentials in immature rat CA3 hippocampal neurones. J Physiol 1989; 416:303-325.
13. Lerma J. Roles and rules of kainate receptors in synaptic transmission. Nat Rev Neurosci 2003; 4:481-495.
14. Lerma J. Kainate receptor physiology. Curr Opin Pharmacol 2006; 6:89-97.
15. Pinheiro P, Mulle C. Kainate receptors. Cell Tissue Res 2006; 326:457-482.
16. Vignes M, Collingridge GL. The synaptic activation of kainate receptors. Nature 1997; 388:179-182.
17. Castillo PE, Malenka RC, Nicoll RA. Kainate receptors mediate a slow postsynaptic current in hippocampal CA3 neurons. Nature 1997; 388:182-186.
18. Lauri SE, Bortolotto ZA, Bleakman D et al. A critical role of a facilitatory presynaptic kainate receptor in mossy fiber LTP. Neuron 2001; 32:697-709.
19. Lauri SE, Segerstråle M, Vesikansa A et al. Endogenous activation of kainate receptors inhibits glutamatergic transmission and modulates network activity in the developing hippocampus. J. Neurosci 2005; 25(18):4473-4484.
20. Schmitz D, Mellor J, Nicoll RA. Presynaptic kainate receptor mediation of frequency facilitation at hippocampal mossy fiber synapses. Science 2001: 291:1972-1976.
21. Melyan Z, Wheal HV, Lancaster B. Metabotropic-mediated kainate receptor regulation of IsAHP and excitability in pyramidal cells. Neuron 2002; 34(1):107-114.
22. Ruiz A, Sachidhanandam S, Utvik JK et al. Distinct subunits in heteromeric kainate receptors mediate ionotropic and metabotropic function at hippocampal mossy fiber synapses. J Neurosci 2005; 25(50):11710-11718.
23. Fisahn A, Contractor A, Traub RD et al. Distinct roles for the kainate receptor subunits GluR5 and GluR6 in kainate-induced hippocampal gamma oscillations. J Neurosci 2004; 24:9658-9668.
24. Fisahn A. Kainate receptors and rhythmic activity in neuronal networks: hippocampal gamma oscillations as a tool. J Physiol 2005; 562(Pt 1):65-72.
25. Nadler JV. Kainic acid as a tool for the study of temporal lobe epilepsy. Life Sci 1981; 29:2031-2042.
26. Ben-Ari Y, Cossart R. Kainate, a double agent that generates seizures: two decades of progress. Trends Neurosci 2000; 23(11):580-587.
27. Lerma J, Paternain AV, Naranjo JR et al. Functional kainate-selective glutamate receptors in cultured hippocampal neurons. Proc Natl Acad Sci 1993; 90(24):11688-11692.
28. Bleakman D, Ogden A-M, Ornstein PL et al. Pharmacological characterization of a GluR6 kainate receptor in cultured hippocampal neurons. Eur J Pharmacol 1999; 378:331-337.
29. Wilding TJ, Huettner JE. Activation and desensitization of hippocampal kainate receptors. J Neurosci 1997; 17(8):2713-2721.
30. Wong LA, Mayer ML, Jane DE et al. Willardiines differentiate agonist binding sites for kainate-versus AMPA-preferring glutamate receptors in DRG and hippocampal neurons. J Neurosci 1994; 14(6):3881-3897.
31. Mulle C, Sailer A, Pérez-Otaño I et al. Altered synaptic physiology and reduced susceptibility to kainate-induced seizures in GluR6-deficient mice. Nature 1998; 392:601-605.
32. Smolders I, Bortolotto ZA, Clarke VR et al. Antagonists of GLU(K5)-containing kainate receptors prevent pilocarpine-induced limbic seizures. Nat Neurosci 2002; 5:796-804.
33. Rogawski MA, Gryder D, Castaneda D et al. GluR5 kainate receptors, seizures and the amygdale. Ann NY Acad Sci 2003; 985:150-162.
34. Kaminski RM, Banerjee M, Rogawski MA. Topiramate selectively protects against seizures induced by ATPA, a GluR5 kainate receptor agonist. Neuropharmacology 2004; 46:1097-1104.

35. Huxter JR, Zinyuk LE, Roloff EL et al. Inhibition of kainate receptors reduces the frequency of hippocampal theta oscillations. J Neurosci 2007; 27(9):2212-2223.
36. Traub RD, Bibbig A, LeBeau FE et al. Cellular mechanisms of neuronal population oscillations in the hippocampus in vitro. Annu Rev Neurosci 2004; 27:247-278.
37. Brown JT, Teriakidis A, Randall AD. A pharmacological investigation of the role of GLUK5-containing receptors in kainate-driven hippocampal gamma band oscillations. Neuropharmacology 2006; 50:47-56.
38. Christensen JK, Paternain AV, Selak S et al. A mosaic of functional kainate receptors in hippocampal interneurons. J Neurosci 2004; 24:8986-8993.
39. Jaskolski F, Coussen F, Nagarajan N et al. Subunit composition and alternative splicing regulate membrane delivery of kainate receptors. J Neurosci 2004; 24:2506-2515.
40. Wang XJ, Buzsáki G. Gamma oscillation by synaptic inhibition in a hippocampal interneuronal network model. J Neurosci 1996; 16:6402-6413.
41. Bartos M, Vida I, Jonas P. Synaptic mechanisms of synchronized gamma oscillations in inhibitory interneuron networks. Nat Rev Neurosci 2007; 8:45-56.
42. Mann EO, Paulsen O. Role of GABAergic inhibition in hippocampal network oscillations. Trends Neurosci 2007; 30:343-349.
43. Cossart R, Esclapez M, Hirsch JC et al. GluR5 kainate receptor activation in interneurons increases tonic inhibition of pyramidal cells. Nat Neurosci 1998; 1: 470-478.
44. Frerking M, Malenka RC, Nicoll RA. Synaptic activation of kainate receptors on hippocampal interneurons. Nat Neurosci 1998; 1:479-486.
45. Bureau I, Bischoff S, Heinemann SF et al. Kainate receptor-mediated responses in the CA1 field of wild-type and GluR6-deficient mice. J Neurosci 1999; 19:653-63.
46. Mulle C, Sailer A, Swanson GT et al. Subunit composition of kainate receptors in hippocampal interneurons. Neuron 2000; 28:475-484.
47. Semyanov A, Kullmann DM. Kainate receptor-dependent axonal depolarization and action potential initiation in interneurons. Nat Neurosci 2001; 4:718-723.
48. Goldin M, Epsztein J, Jorquera I et al. Synaptic kainate receptors tune oriens-lacunosum moleculare interneurons to operate at theta frequency. J Neurosci 2007; 27:9560-9572.
49. Yang EJ, Harris AZ, Pettit DL. Synaptic kainate currents reset interneuron firing phase J. Physiol 2007; 578:259-273.
50. Paternain AV, Herrera MT, Nieto MA et al. GluR5 and GluR6 kainate receptor subunits coexist in hippocampal neurons and coassemble to form functional receptors. J Neurosci 2000; 20:196-205.
51. Henze DA, Wittner L, Buzsáki G. Single granule cells reliably discharge targets in the hippocampal CA3 network in vivo. Nat Neurosci 2002; 5:790-795.
52. Mori M, Abegg MH, Gähwiler BH et al. A frequency-dependent switch from inhibition to excitation in a hippocampal unitary circuit. Nature 2004; 431:453-456.
53. Bahn S, Volk B, Wisden W. Kainate receptor gene expression in the developing rat brain. J Neurosci 1994; 14:5525-5547.
54. Ritter LM, Vazquez DM, Meador-Woodruff JH. Ontogeny of ionotropic glutamate receptor subunit expression in the rat hippocampus. Brain Res Dev Brain Res 2002; 139:227-236.
55. Kidd FL, Isaac JT. Developmental and activity-dependent regulation of kainate receptors at thalamocortical synapses. Nature 1999; 400:569-573.
56. Kidd FL, Coumis U, Collingridge GL et al. A presynaptic kainate receptor is involved in regulating the dynamic properties of thalamocortical synapses during development. Neuron 2002; 34:635-646.
57. Lee CJ, Kong H, Manzini MC et al. Kainate receptors expressed by a subpopulation of developing nociceptors rapidly switch from high to low Ca2+ permeability. J Neurosci 2001; 21:4572-4581.
58. Lauri SE, Vesikansa A, Segerstråle M et al. Functional maturation of CA1 synapses involves activity-dependent loss of tonic kainate receptor-mediated inhibition of glutamate release. Neuron 2006; 50:415-429.
59. Sallert M, Malkki H, Segerstråle M et al. Effects of the kainate receptor agonist ATPA on glutamatergic synaptic transmission and plasticity during early postnatal development. Neuropharmacology 2007; 52:1354-1365.
60. Vignes M, Clarke VR, Parry MJ et al. The GluR5 subtype of kainate receptor regulates excitatory synaptic transmission in areas CA1 and CA3 of the rat hippocampus. Neuropharmacology 1998; 37(10-11):1269-1277.
61. Clarke VRJ, Collingridge GL. Characterisation of the effects of ATPA, a GLUK5 receptor selective agonist, on excitatory synaptic transmission in area CA1 of rat hippocampal slices. Neuropharmacology 2002; 42:889-902.
62. Partovi D, Frerking M. Presynaptic inhibition by kainate receptors converges mechanistically with presynaptic inhibition by adenosine and GABAB receptors. Neuropharmacology 2006; 51:1030-1037.

63. Maingret F, Lauri S, Taira T et al. Profound regulation of neonatal CA1 rat hippocampal GABAergic transmission by functionally distinct kainate receptor populations. J Physiol 2005; 567(Pt 1):131-142.
64. Rodriguez-Moreno A, Herreras O, Lerma J. Kainate receptors presynaptically downregulate GABAergic inhibition in the rat hippocampus. Neuron 1997; 9:893-901.
65. Jiang L, Xu J, Nedergaard M et al. A kainate receptor increases the efficacy of GABAergic synapses. Neuron 2001; 30:503-513.
66. Garaschuk O, Hanse E, Konnerth A. Developmental profile and synaptic origin of early network oscillations in the CA1 region of rat neonatal hippocampus. J Physiol 1998; 507:219-236.
67. Khalilov I, Dzhala V, Medina I et al. Maturation of kainate-induced epileptiform activities in interconnected intact neonatal limbic structures in vitro. Eur J Neurosci 1999; 11:3468-3480.
68. Tremblay E, Nitecka L, Berger ML et al. Maturation of kainic acid seizure-brain damage syndrome in the rat. I. Clinical, electrographic and metabolic observations. Neuroscience 1984; 13:1051-1072.
69. Khalilov I, Hirsch J, Cossart R et al. Paradoxical anti-epileptic effects of a GluR5 agonist of kainate receptors. J Neurophysiol 2002; 88:523-527.
70. Hennou S, Khalilov I, Diabira D et al. Early sequential formation of functional GABA(A) and glutamatergic synapses on CA1 interneurons of the rat foetal hippocampus. Eur J Neurosci 2002; 16:197-208.
71. Tashiro A, Dunaevsky A, Blazeski R et al. Bidirectional regulation of hippocampal mossy fiber filopodial motility by kainate receptors: A two step model of synaptogenesis. Neuron 2003; 38:773-784.
72. Marchal C, Mulle C. Postnatal maturation of mossy fibre excitatory transmission in mouse CA3 pyramidal cells: a potential role for kainate receptors. J Physiol 2004; 561(Pt 1):27-37.
73. Vesikansa A, Sallert M, Taira T et al. Long-term activation of kainate receptor subunit GluR5 increases mEPSC frequency in the hippocampal slice cultures. J Physiol 2006; 583(Pt 1):145-157.
74. Huupponen J, Molchanova SM, Taira T et al. Susceptibility for homeostatic plasticity is down-regulated in parallel with maturation of the rat hippocampal synaptic circuitry. J Physiol 2007; 581(Pt 2):505-514.

Chapter 9

Kainate Receptor Modulation by Sodium and Chloride

Andrew J.R. Plested*

Abstract

The kainate-type glutamate receptor displays strong modulation by monovalent anions and cations. This modulation is independent of permeation of the ion channel. Instead, structural, computational and biophysical evidence shows that receptor activity is controlled by binding of sodium and chloride ions at sites that stabilize active dimers of glutamate binding domains. Modulation by monovalent ions is a surprisingly general property across ion channel families. However, evidence of a physiological role for ion-dependent effects on glutamate receptors is lacking, perhaps reflecting the adventitious use of ions as structural components of the kainate receptor.

> *"ergo, Hercules, vita humanior sine sale non quit degree [...]"*
> *"Heaven known, a civilized life is impossible without salt"*
> *—Pliny the Elder, Natural History XXXI 88*

Introduction

The essential nature of salt in biology, over and above it's role in civilized human life, cannot be understated. As biologists, our appreciation of sodium and chloride dates to the investigations of Sydney Ringer (for a stimulating discussion of Ringer's work, see ref. 1). Serum salt concentration is remarkably invariant, because peripheral reflexes are employed to keep it so.[2] We might then expect sodium and chloride ions to be inert, having no modulatory roles in protein function and think it odd that any protein would bind these ions in a key functional region. But sodium and chloride modulate many proteins, including Thrombin and Hemoglobin,[3,4] acting at well-defined sites. In the same way, some ion channels, in addition to passing sodium and chloride across the cell membrane, are also subject to modulation by these ions.

Physical Characteristics of Sodium and Chloride

An excellent survey of the physical properties of ions is available.[5] Water molecules (55 M) vastly outnumber sodium and chloride ions (each ~140 mM) in the cerebral-spinal fluid. Thus, water can cluster around each ion and shield it, because the electric field of the ion polarizes the dipoles of the bound waters. For this reason, dehydrating the ion requires energy and the coordinating groups of any proteinaceous binding site must provide this energy. Energies of hydration are high, but the sphere of water molecules and the stoichiometry of coordination, are not fixed, but instead vary on the nanosecond timescale. How can such loose behaviour be consistent with selective binding of ions to protein sites? Exquisite selection can be achieved by providing a coordination environment that differs, serially according to ion chemistry, from that given by bulk water.[6]

*Andrew J.R. Plested—Leibniz-Institut für Molekulare Pharmakologie (FMP), Robert-Rössle-Str. 10, 13125 Berlin, Germany. Email: plested@fmp-berlin.de

Kainate Receptors: Novel Signaling Insights, edited by Antonio Rodríguez-Moreno and Talvinder S. Sihra.

Sodium and chloride ions are both singly charged, but have quite different personalities. Asymmetry immediately arises from the electron. Chloride ions harbor an extra electron, but its negative charge is thinly spread over the surface. This renders chloride ions rather sticky and prone to bind indiscriminately to protein surfaces. Crystallographers exploit this propensity to bind in the form of halide cryosoaking.[7] In this method, bound chloride ions produce anomalous diffraction signals, which can be used to obtain phase information essential for protein structure solution. This kind of anomalous signal is also useful in fingerprinting bound halides in crystal structures (see below). Sodium, by contrast, is much more compact. By losing an electron, it gains a charge but becomes smaller and this adds up to a fairly high surface charge density. Oxygen ligands are thus drawn closer, with characteristic metal-ligand distances that are significantly shorter than hydrogen bonds (on average 2.4Å).[8] The rapid exchange of ligands in bulk water is mirrored in proteinaceous binding sites, where sodium ions might be comfortably coordinated by between 3 and 7 ligands[8,9] and have similar thermal motion to the surrounding protein.[10] A technical problem with the detection of sodium ions in crystallographic experiments is that they have the same number of electrons as water. Thus, valence screening, which relies on bond distances, rather than any specific coordination number or scattering measurements, is the best method for de novo identification, at least in high-resolution structures where coordinate errors are small.[11]

When we consider ion binding to proteins, the chemistry of amino acids gives a second layer of asymmetry. Chloride ion binding sites regularly include a basic side chain, either Arg or Lys. These moieties are comparatively long and hence can sweep out large swathes of space if not restrained. By contrast, cation binding sites can include negatively-charged carboxyl groups of Asp and Glu, to counterbalance positive charge on the cation. Simple turn-loop motifs are common in divalent cation binding sites,[12] possibly because these donors are compact. The kainate receptor monovalent cation binding sites follow such a pattern of utilizing groups from adjacent amino acids in a turn region.

Structural View of Glutamate Receptor Activation

Before discussing the ion binding sites of kainate receptors in detail, it is instructive to consider the overall structure and gating of glutamate receptors. Reviews on this broader topic are available.[13,14] Our current view of the glutamate receptor is defined by the structure of the GluA2 tetramer (Fig. 1).[15] This crystal structure is probably also an excellent model of the tertiary structure of kainate receptors.[16] The overall structure of GluA2 is also consistent with the long held idea that the four ligand binding domains each close on the agonist to open the channel. Generally, the more ligand binding domains that are occupied,[17] or the more profound the closure of each clamshell domain,[18] the stronger the activation of the channel will be.

The channel pore of glutamate receptors conducts cations and is similar in architecture to other tetraspanning channels. A reentrant membrane loop forms a narrow constriction that selects permeant ions. At the tip of this loop, the genomically-encoded Glutamine can be RNA-edited to Arginine and this editing exerts strong control over ions transiting the pore.[19] In AMPA and kainate receptors, if all subunits harbour glutamines, the pore is equally well permeated by all monovalent cations up to Cs and also passes Ca^{2+}.[20,21] This information is important in the context of allosteric modulation by other ions than sodium—in kainate receptors, monovalent ions have effects on permeation that are minimal. When Arg residues are present, as in the majority of native AMPA receptors that contain the edited GluA2 subunit and in kainate receptors that contain edited GluK1 or GluK2 subunits,[22] the single channel conductance is decimated and anions such as F^- are permeant.[21]

Each subunit has a modular structure, but the protein sequence is convoluted, because the channel domain is interdigitated with the ancient bilobed ligand binding domain.[23] A similar clamshell domain sits N-terminal to the glutamate binding core (amino terminal domain, ATD) and determines subtype-specific assembly.[24] In neurons, these domains distal to the pore are also likely to bind proteins of the extracellular matrix and so influence synaptic activity by controlling

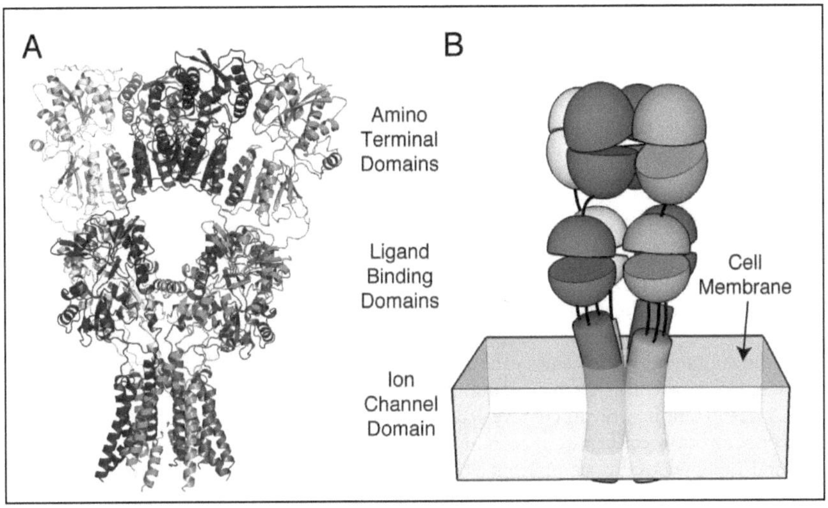

Figure 1. Glutamate receptor structure. A) The resting state crystal structure of the GluA2 tetramer (PDB code: 3KG2)[15] is likely to be a good model for AMPA, kainate and NMDA-type glutamate receptors. The four subunits are coloured red, green, blue and yellow. The extracellular clamshell domains are loosely coupled by flexible linkers and form dimers with alternating partner subunits. B) A simplified cartoon schematic of the domain organization in glutamate receptor, with colouring as in panel A. Multiple intersubunit contacts in the extracellular doman control receptor function. When glutamate binds to the open clamshell Ligand Binding Domains (LBDs) the clamshells close and drive opening of the membrane spanning ion channel pore. The dimer pairs of the LBDs form the binding sites for allosteric ions in kainate receptors (Blue/Yellow and Red/Green subunits). A color version of this figure is available at www.landesbioscience.com/curie.

receptor mobility.[25,26] Local dimers, like those previously determined from the crystal packing of isolated ligand binding domains (LBDs) and ATDs of kainate receptors,[27,28] are well preserved in the full-length GluA2 structure (see Fig. 1). The conservation of dimeric interactions between the extracellular domains is particularly important when considering the binding sites of allosteric monovalent ions, because these sites are integral structural components of the exquisitely-tuned dimer interaction between the ligand binding cores. The dimers form two "layers". The upper ATD layer arranged as an extended N-shaped tetramer[15,16,29] and the LBDs are more compactly packed in a four-square arrangement immediately extracellular to the membrane. Disulfide crosslinks based on the GluA2 receptor antagonist-bound structure also form in GluK2 kainate receptors and between GluN2 NMDA subunits.[15,16] Therefore, to a first approximation, the arrangement of the extracellular domains and their relation to the pore domain is the same in AMPA, NMDA and kainate receptors.

The iGluR ion channel has a tetrameric symmetry and the membrane probably has a stabilizing role at interfaces between subunits at this level. The intersubunit interactions between extracellular domains have a spectrum of stabilities, ranging from the undetectably weak (the dimer-dimer interfaces), to the tight (the dimer interfaces between the ATDs of GluK2; dissociation constant about 15 μM; ref. 27). The affinity of the ligand binding core dimers that accumulated evidence shows are also present in the active state, have intermediate dissociation constants (~5 mM). This association is strong enough to influence crystal packing in the right conditions but generally too weak to detect between isolated binding sites in solution. This balance can be shifted towards dimerization by targeted modifications and also by the binding of exogenous compounds. For the ATDs, the dimer dissociation constants are much lower and so these domains (when dimerized) are thought to be much less dynamic than the ligand binding dimers. Deletion of the ATD is shown to leave AMPA and kainate activation unaffected,[30,31] suggesting a nonessential role in channel gating. However, disulfide mutants in the ATD layer perturb gating upon oxidation,[16] although the functional effect of restraint is more modest than that of crosslinking the ligand binding cores. In NMDA receptors, control of channel open probability by signaling from the ATDs is well-established.[32,33] In both cases the ATD probably exerts control over the gating of the pore by altering the tetrameric packing of the LBD layer.

Ion Binding Sites in the Dimer Interface of Kainate Receptors

In the absence of structural data, insight to the nature of ligand binding sites can be obtained from site-directed mutagenesis (for example see ref. 34). Unfortunately, mutagenesis is a blunt tool when investigating the small, highly-charged cavities where ions bind. Protein side chains sample limited chemical space and backbone carbonyls (that are inaccessible to mutagenesis) participate in coordination. To identify the location of the ion binding sites in kainate receptors and investigate their chemistry, structural information was essential. High-resolution structures of the GluK1 ligand binding core dimer revealed bound chloride ions on the two-fold dimer axis.[31] This was unequivocally confirmed in bromide-soaked crystals of GluK1 (PDB code: 2OJT). In these crystals, anomalous diffraction from a bound bromide ion (a functional surrogate of chloride) was observed at the anion site. The dimerization interface where chloride binds is formed between the backs of the upper lobes (D1) of the ligand binding domains. In AMPA receptors, this dimer interface was already well understood to harbor binding sites for exogenous positive modulators of desensitization.[35,36] Thus, it was logical to suspect that chloride bound at this site could modulate kainate receptor desensitization.

Sodium or chloride ions are not resolved at or near to the D1 sites in any of the monomeric crystal structures of GluK1 or GluK2 (e.g., PDB code 1S50).[37] However, packing of different crystal forms mandates that such observations are interpreted with extreme caution. Although head-to-tail dimers of kainate receptors can be generated from some structures, this arrangement is completely nonphysiological with relation to the full-length receptor. Furthermore, in some dimer structures, the cation sites are unexpectedly plugged by Lys side chains, again from LBDs apposed in a nonphysiological contact, driven purely by crystal packing. This is the case in the GluK1-UBP302 complex (PDB 2F35).[38] Crystallization experiments using kainic acid, in order to sample a larger conformational space for crystal packing, succeeded in freeing the cation sites from blockade by Lysine and allowed occupancy of the cation site by a range of monovalent ions.[39] These kainate-bound structures retained chloride ions bound in the same dimer site flanked by a pair of cations, at least in the presence of Li^+ and Na^+ ions (see below).

Anion sites were originally described in dimers of the GluK1 receptor ligand binding cores that were stabilized by antagonist molecules (Fig. 2). The domain 1 (D1) dimer interface was intact, but this structure most closely represented a resting state. For Na and Cl ions to slow desensitization, one would expect them to stabilize agonist-bound dimers in the same active conformation. Satisfyingly, the arrangement of chloride and sodium ions is well preserved in dimers of GluK1 to which the partial agonist kainate is bound[39] and is likewise similar in a set of crystal structures of wild-type and mutant GluK2 LBDs in the active dimer arrangement with glutamate bound[40,41] (Fig. 3). Structural information about the ligand binding cores in the absence of agonist or antagonist molecules and the desensitized arrangement of the dimer interface, is not currently available for kainate receptors. Thus, the status of the ion binding sites in these states remains an open question. The interactions that underlie chloride and sodium binding in the wild-type GluK2 dimer are shown in Figures 4 and 5.

Structural models of the ligand binding cores with ions bound made sense of some of the existing site-directed mutagenesis, insofar as disruption of residues that coordinate chloride generally sped desensitization[42] presumably by increasing the chloride dissociation constant and thus allowing balancing charge to escape more quickly from the interface. The importance of the ions as structural components of the active interface underlines the common dimer arrangement of ligand binding

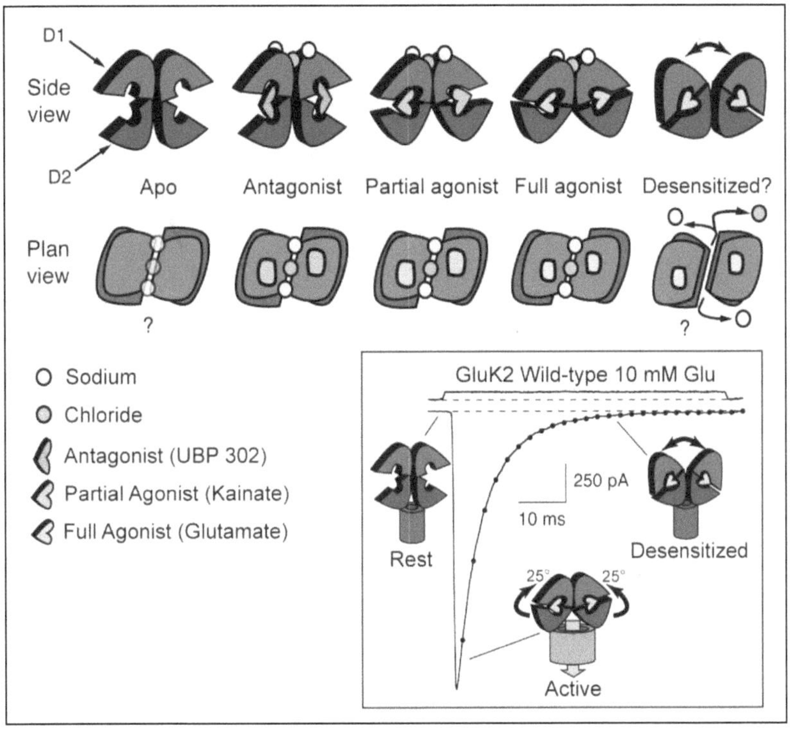

Figure 2. Cartoon representations of state-dependent ion binding to kainate receptor ligand-binding domain dimers. Monovalent ions are resolved in high-resolution crystal structures for antagonist, partial agonist and full agonist complexes (GluK1-UBP302, PDB: 2F35; GluK1-Kainate, PDB: 3C32 and GluK2-Glutamate, PDB: 3G3F). The ions are found at the interface between the upper lobes, with chloride sandwiched between sodium ions on the two-fold axis. No apo or desensitized structures are available for kainate receptors. Published candidate structures for AMPA receptors (GluA2 apo, PDB: 1FTO and GluA2 S729C desensitized dimer, PDB: 2I3W) suggest the D1 dimer interface is intact in the apo state and ruptured in the desensitized state. Here, the probable occupancy of the ion sites is indicated accordingly (see text). Boxed inset shows a simplified scheme of kainate receptor activation. Glutamate binding drives a conformational change that opens the channel. This is quickly followed by dissociation of the active dimer, which relaxes in to the nonconducting desensitized state. Thus ions that fail to stabilize the dimer interface speed desensitization. A color version of this figure is available at www.landesbioscience.com/curie.

cores in all members of the superfamily. AMPA receptors are insensitive to the binding of ions, because they lack dimer interfacial sites to coordinate chloride and because typically Lys residues plug the cation sites (discussed in ref. 39).

Additional ion sites on the extracellular side, outside the electric field, cannot currently be ruled out but must be considerably weaker in influence than those in the D1 dimer interface. Removal of the ATD has no effect on ion modulation,[31,39] suggesting that this region has no functionally significant binding sites for monovalent ions. Point mutants in the cation sites can completely remove sensitivity. This is not the case for chloride binding site mutants, which might be taken to imply additional chloride sites. But these results could also easily be due to coupling between the cation and anion binding sites and the bluntness of site directed mutagenesis, rather than

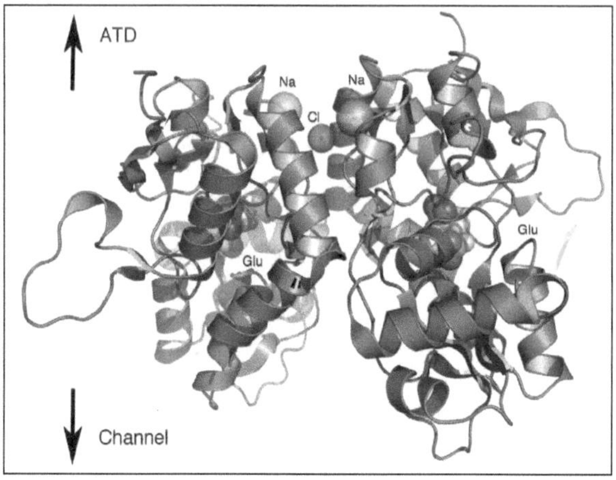

Figure 3. Sodium and chloride ions stabilize the active LBD dimer. High-resolution crystal structures of the isolated ligand binding core of GluK2 in complex with glutamate (PDB code 3G3F) reveal ion binding sites in the dimer interface. Chloride is bound in the same site in antagonist and partial agonist complexes. One of the two dimers in the full-length tetrameric receptor is shown, from the side. Glutamate molecules bind in the jaws of the clamshells. Each subunit has a Na^+ binding site and these flank the single Cl^- ion. The Na^+ and Cl^- ions are bound in the upper reaches of the interface between domains 1, stabilizing the active dimer conformation. A color version of this figure is available at www.landesbioscience.com/curie.

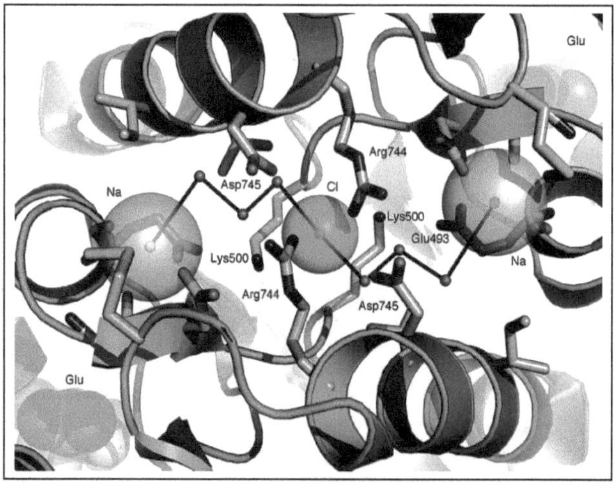

Figure 4. Coupling between ions in the dimer interface. Plan view of the active D1 dimer interface, illustrating coupling between the sodium (gold spheres) and chloride ions (green sphere). Twin salt bridges across the GluK2 D1 dimer interface between Arg 744 and Asp745 form the roof of the anion site. A highly-ordered chain of 3 water molecules (red spheres linked by black lines) links each sodium ion to the chloride ion bound at the two-fold axis. A parallel connection is formed by the carboxyl groups of Glu 493, which coordinate each sodium ion and form salt-bridges across the dimer interface to Lys 500, whose ε-amino groups balance negative charge at the base of the anion site. A color version of this figure is available at www.landesbioscience.com/curie.

additional chloride sites. Negatively-charged side chains in the cation site can be exchanged for isosteric polar side chains (Asp to Asn, for example), but the positively charged groups that form the anion site have no polar analogues of similar size.

The Role of Ions in Kainate Receptor Gating

Before cloning of the separate glutamate receptor subtypes, experiments in native neurons with nonselective agonists demonstrated that kainate-activated currents were reduced by chloride substitution.[43] When the GluK2 subunit was initially cloned, electrophysiological experiments to determine ion permeability revealed a similar inhibition of the current when Chloride was replaced by methanesulfonate.[44]

Both these findings were consistent with the allosteric inhibition found later, but the conscious identification of ion-sensitive activation of kainate receptors arose serendipitously from experiments to measure recovery from desensitization of AMPA and kainate receptors.[45] These experiments were done with fast perfusion on cloned receptors and so had the twin advantages that the molecular nature of the target was known and receptors were activated by glutamate with kinetics similar to synaptic currents. To boost the small agonist-activated currents during the early phase of recovery, experiments were performed with hypertonic saline. This manipulation revealed a kinetic distinction between GluA1 and GluK2 receptors. The rate of recovery from desensitization was affected by the sodium chloride concentration, but only for GluK2. Subsequent experiments revealed large differences in the size and kinetics of kainate receptor responses to glutamate in the presence of different ionic species.[46] Strikingly, both cations and anions are able to exert very strong effects on the desensitization rate for GluK2, but AMPA receptors are largely spared. This phenomenon is illustrated in Figure 6 for glutamate-activated currents.

These pioneering experiments on GluA1 and GluK2 immediately excluded several mechanisms of modulation. Firstly, no effect was observed from intracellular ion substitutions and the inhibition from ions other than sodium and chloride was voltage-insensitive. So, even though all monovalent cations up to Cs^+ can permeate the pore, permeating ions are not responsible for changes in gating. These studies pointed to binding sites for ions in the extracellular domains, or the linkers between the channel and the binding sites, where for example noncompetitive antagonists can bind.[47] Another important observation was that the slowing of the decay in high-ionic strength solutions was broadly in opposition to competitive antagonism by ions at the ligand binding site.

Chimeras formed from GluA3 and GluK2 receptors were employed to identify an amino acid in the kainate receptor ligand binding core (Met 739 in GluK2) where Cs^+ inhibition could be effectively nullified by substitution with the Lysine residue found in this position in AMPA receptors.[48] It is now clear, that at least in GluA2, this Lysine can plug the cation site. A second residue, Asp 497, invariant between kainate receptors and the relatively-insensitive AMPA receptors could also be truncated to Ala in order to remove Cs^+ inhibition. Further interpretation was also hampered by the failure of the converse mutation at the Met site to induce cation sensitivity in GluA3. This mutant was only functional in cyclothiazide, well-characterized as a blocker of AMPA receptor desensitization. Given what is now known about the mechanism of ion modulation, even if this mutant were ion-sensitive, it is probable that cyclothiazide would have obscured ion modulation, by preventing dimer dissociation.

Bound Ions Control the Biophysical Properties of Kainate Receptors

How do ions alter kainate receptor function? Ion substitutions that inhibit maximum currents also speed desensitization. Measurements of glutamate activated currents in a range of sodium and chloride concentrations and at different ion strengths reveal coupled modulation, with reciprocal control of affinity. Sodium binding makes chloride binding tighter, but in physiological salt, the receptor is probably not fully bound by either ion. The affinity for sodium is weaker than 100 mM and the site is half bound by Cl^- at a concentration of 90 mM.[39] How does receptor activity then increase or decrease in different ionic conditions? Naively, one might expect that the conductance of the ion channel, with its own cation binding sites, will be strongly perturbed. But manipulations that both

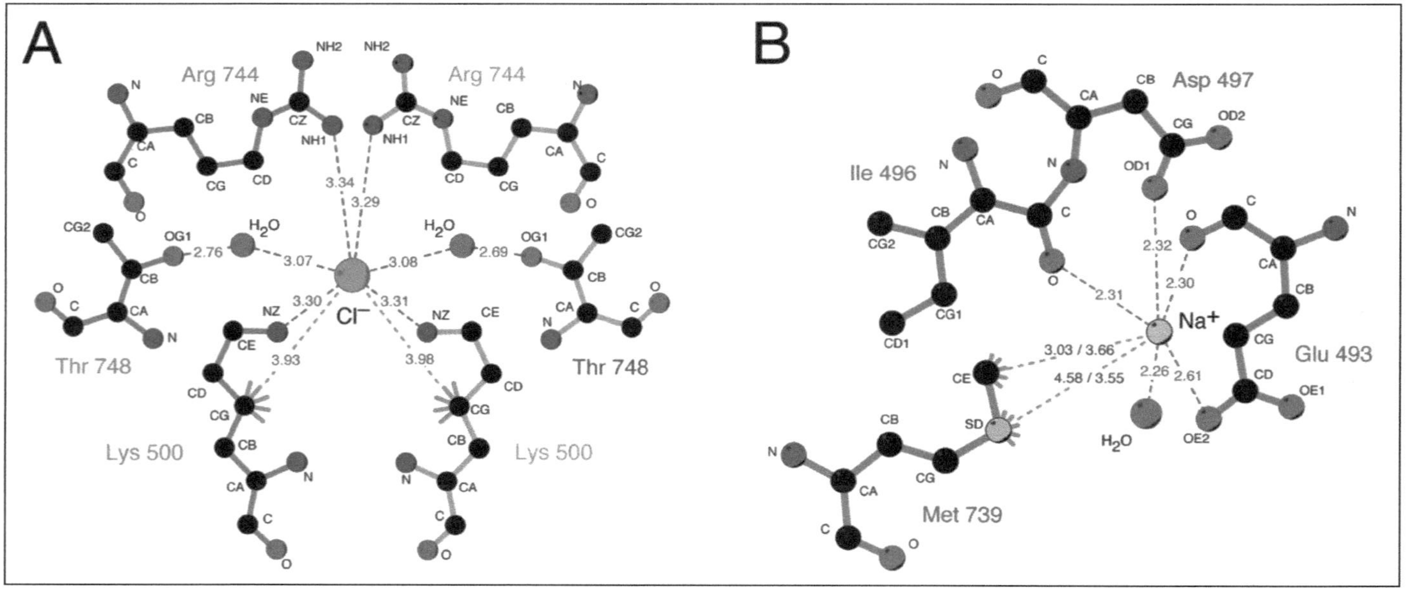

Figure 5. Interactions in the allosteric ion binding sites in GluK2. A) Flattened schematic of the anion binding site in the GluK2-glutamate complex (PDB: 3G3F). Positively charged groups from separate subunits form the chloride site at the dimer two-fold axis. Twin Arginine residues are highly mobile, adopting multiple conformations. The down conformation is shown above. When engaged with the anion, the guanidinium groups of Arg 744 also participate in intersubunit salt bridges. Twin Lysine residues are comparatively static and also make salt-bridge interactions across the dimer interface, with Glu 493, as well as a probable Van-der-Waals contact with the Chloride ion. Twin water molecules bridge the chloride ion to Thr 748 and to the sodium ion *via* a water chain. B) Sodium sites from the GluK2-glutamate dimer are solvent exposed and entirely formed from single subunits. Met 739 adopts two conformations, both are shown. Either the terminal Methyl group caps one aspect of the cation site, or both the Methyl group and the sulphur atom bolster the bound sodium ion *via* Van-der-Waals interactions. Metal-oxygen distances are characteristically short for the bonds to both the main chain carbonyls and for the monodentate coordination by negatively charged carboxyl groups of Asp 493 and Glu 497. A color version of this figure is available at www.landesbioscience.com/curie.

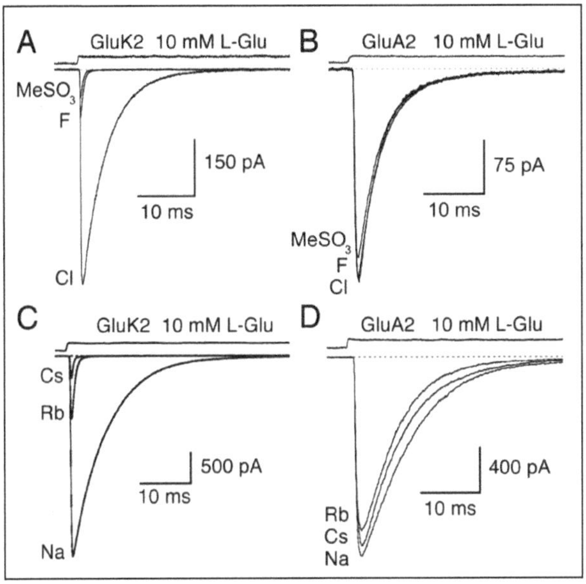

Figure 6. Kainate receptor currents are strongly sensitive to ions. A) Outside-out patches containing glutamate receptors expressed in HEK 293 cells were subjected to fast solution exchange. Wild-type kainate receptors activated by a brief pulse of 10 mM Glutamate are strongly inhibited when chloride is replaced by either fluoride or methanesulfonate on a sodium background. When chloride is substituted, the desensitization rate is accelerated nearly 10-fold. Salt concentration was 150 mM in each case. B) The same ion exchanges have no effect on wild-type AMPA receptors. C) Strong inhibition of kainate receptor currents and speeding of desensitization also occurs when sodium is substituted by Cesium or Rubidium on a chloride background. D) Cation exchanges have minimal effects on wild-type AMPA receptor currents, even though the permeant ion is swapped. Modified from reference 39.

increase and decrease receptor activity have remarkably little effect on the weighted single channel conductance[31,39] and chloride is barely permeant. Noise analysis of macroscopic currents instead shows that the main effects of allosteric ions are to alter peak open probability and the number of active receptors. A reduction in the number of active receptors when chloride is replaced by fluoride might reflect a greater proportion of receptors with the D1 interface disrupted at rest and therefore not competent for activation by glutamate. Spectroscopic measurements and state-dependent cross-linking of AMPA channels suggest that the D1 dimer interface can dissociate at rest, consistent with this idea.[49,50] In line with the nonsaturation of the ion binding sites in physiological conditions, the number of active receptors can also be increased, by raising NaCl concentration. In the context of the full-length receptor, ion-dependent changes in open probability might arise from rigid body rearrangements of the ligand binding cores. The ion binding sites are located at the very top of the D1 interface (see Fig. 3) and this might cause the bound LBDs to act on the ion channel as 50 Å-long levers. For example, in the case of strong inhibition by fluoride,[46] the braced dimer arrangement might relax, in order to accommodate the extreme electrostatic requirements of fluoride ions. If domains 1 move marginally apart, domains 2 would tend to close together; that is, they would approach the resting state configuration, but without a change in agonist-induced domain closure. Thus the kainate receptor ion channel would be activated less. In support of this model, crystal structures with cations bound in the presence of chloride had tighter dimer arrangement than antagonist-bound chloride complexes.[39] However, in the same study, the dimer arrangement with less efficacious cations bound was shown to be the same as that in sodium. Instead of a change in tilt angles between LBDs, bound chloride became progressively disordered in Rb^+ and Cs^+ complexes. However, in these cation soaking experiments, it is possible that crystal contacts overcame weak changes in the dimer tilt angles accompanying binding of less efficacious cations, which would otherwise be translated into reduced activation in the full length channel.

Ions and Desensitization

Upon binding of glutamate and the subsequent complex transition to the open state, the D1 dimer interface is likely to experience both perpendicular strain and torsion. It is this force that is thought to break the D1 dimer interface, leading to desensitization. To date, no desensitized state structure has been determined for the kainate receptor. We rely on analogy with the Cys-linked dimer of GluA2 ligand binding cores, which has structural and pharmacological properties suggesting a glutamate-bound, inactive state.[50,51] In this structure, the lower lobes (D2) of the ligand binding cores have moved together, consistent with a closed channel, and the upper lobes have dissociated. If the conformational changes are similar in the kainate receptor, ion binding sites in the interface between the upper lobes of the LBD are presumably disrupted upon desensitization (Fig. 2). Whilst such a structure for kainate receptors might give insight to the disruption of the ion binding sites when the D1 interface is broken, several pieces of evidence suggest that these ideas apply to the desensitized kainate receptor. Most importantly, entry to desensitization is a within-dimer phenomenon.[52] Although the final, very stable, kainate receptor desensitized state may involve tetramer rearrangements, this state cannot be accessed by crosslinked dimers and these covalent dimers exhibit no ion sensitivity.[31,39] These dimers can be activated normally in nominally 'ion-free' solutions that contain only sucrose and buffer molecules, titrated to pH 7.4. Mutations in the D1 interface and ion exchanges also have very little impact on recovery from desensitization—at most an order of magnitude less than the effects on entry to desensitization.[31,39]

The dimer interface of AMPA receptors harbors a site for a very potent mutation that effectively blocks dimer dissociation (L-Y mutation).[53] Insertion of a Tyrosine at position 483 in GluA2 (or at equivalent positions in other AMPA receptors) increases dimer affinity ~10^5-fold.[36] This mutation was found from chimeras of AMPA and kainate receptors, although in kainate receptors, the presence of Tyrosine does not block desensitization.[53] The reason, at least in part, is that the mutation stabilizes the dimer because Tyrosine docks into a slot in the opposing AMPA receptor subunit and this slot is absent in kainate receptors. Some considerable effort has focused upon engineering the dimer interface to gain a nondesensitizing kainate receptor.[40,52,54] Results in this respect have in general been disappointing, because the very stable desensitized state of GluK2[45] acts as a sink. In the best case, stabilizing the active dimer through the HERLK quintuple mutant slows entry to desensitization nearly 200-fold, but the very slow recovery of wild-type GluK2 means that steady-state desensitization remains substantial (81%).

Surprising insight into the charge balance of the anion binding sites is given by a recent report of an entirely nondesensitizing kainate receptor, where 4 net positive charges were added to the lid of the anion binding site.[55] Aspartate residues that participate in intersubunit salt bridges[31] were swapped for Arginines. In this case, the dimer interface has a low dissociation constant (about 1 μM from AUC measurements), although dimerization of the LBDs is not as tight as for the L483Y mutation in GluA2 (dissociation constant 30 nM).[36] Unfortunately, the D745K mutant could not be crystallized and so the direct influence of chloride on dimerization cannot be taken for granted. If chloride is bound as an essential part of the D1 dimer interface in the D745K mutant, this binding should be much tighter than in wild-type GluK2 (90 mM in physiological salt;).[31] In the full-length D745K mutant, anion and cation sensitivity remains; however, the possibility of very tightly bound contaminant anions cannot be discounted. Certainly, a divalent anion might be expected, from first principles, to provide the appropriate charge balance. However, MD simulations of the D1 dimer interface indicate that a strong electrostatic barrier opposes anion binding when the interface is intact.[56] Additional positive charge in the lid of the anion site would tend to remove this barrier, allowing chloride to bind easily. This situation is reminiscent of the adventitious divalent ion binding site in the GluK2 M739D mutant. Almost all mutants that alter ion sensitivity also speed kainate receptor desensitization. The M739D mutant removes cation sensitivity but slows desensitization.[57] Here, contaminant divalent ions appear to bind to the cation site with much higher affinity than sodium and other monovalent ions. Micromolar concentrations of calcium and magnesium slow desensitization effectively, even in the complete absence of small monovalent cations.[58] Unfortunately, the GluR5 I739D ligand binding core is not expressed as soluble protein and so cannot be subjected to biophysical or structural studies to confirm the presence of bound divalent ions.

The mechanisms of increased dimer interface stabilization by the I739D-Ca and D745K mutants are perhaps related. In both cases, extra charged groups are available to in ion coordination, exceeding the typical count of charged residues in monovalent ion sites.[9] In the case of the pair of basic side chains (Arg744 in GluK2) in the lid of the anion site, crystal structures and molecular dynamic simulations suggest mobility of these side chains, or entropy. Additional positive charge here may permit the chloride ion to lock the side chains in place more securely, thus restraining the dimer effectively by reducing the overall mobility of the interface. Replacing Met 739 with the charged Asp or Glu side chains might equally make the cap of the cation site less mobile when divalent cations are bound, reducing the off-rate compared with sodium and hence producing a higher affinity site.

Physical Basis of Ion Modulation

Despite the setbacks in rational engineering of the kainate receptor dimer interface, mutants with increased dimer stability have been put to good service. The HERLK sites are distant enough from the ion binding sites not to disrupt them and thus ion modulation is identical in the HERLK mutant over the entire range of desensitization rates. This mutant forms a dimer in solution that is stable enough to permit accurate physical determination of the dimer dissociation constant by analytical ultracentrifugation (AUC). Upon this background, the stabilization of the interface by ion binding was measured for the first time.[41] Saturating the ion binding sites was demonstrated to have a similarly strong effect on dimer stability as the five mutations in the interface. The influence of sodium and chloride ions was apparently similar and dimer affinity could also be substantially reduced by replacing sodium and chloride with ions that are negligibly active in functional experiments on wild-type channels ($CsMeSO_3$). The same approach of engineering a more stable interface was extended to the dimer of ∂2 ligand binding domains, which is stabilized by calcium.[59,60] In this case, physiologically relevant calcium concentrations were able to increase the dimer stability, consistent with functional measurements. Thus a common mechanism of allosteric control of glutamate receptors by biological ions, through stabilization of the dimer form, was revealed.

Physical measurements of ion dependent dimer stability and others in this line of study, are clouded by the impossibility of working in the presence of only sodium (in the absence of anions) or chloride (in the absence of cations). The ion binding sites are exposed to the surface of the protein and hence have inherent flexibility, reducing their selectivity by allowing different coordination geometries. Thus, almost all ions retain some activity. The similarity of the functional effects of cation and anion substitutions relates to their common site of action, which gives rise to inevitable coupling between the sites. The close proximity of the sites and the highly-charged surfaces of the binding cavities suggests that a true separation of ion dependent effects between the cation and anion sites may not be achievable.

The Dynamics of Ions in the Dimer Interface

How best can we relate the promotion of association into dimer forms by ions and static high-resolution structural models, to the dynamics of receptor function? Although electrostatic principles may be satisfied in these models, crystal structures cannot yield even simple information about the order or mechanism of ion binding. How do ions enter their sites? The sites are closely apposed and functional measurements clearly indicate coupling. The systems of salt bridges and ordered water molecules that connect the anion and cation sites (Fig. 4) are unlikely to be stable in the absence of one or other of the bound ionic species. Paradoxically, one species might facilitate the binding of the other, although the dimer interface might not form properly without both sodium and chloride ions bound. Therefore, it is a simplification to ignore partly bound states.

Even though such states are undetectable in experiments, they must exist. For a complete picture, one should include 'encounter complexes' involving half-sites and subsequent 'induced fit'.[61] What insights are available from simulations? Unfortunately, one cannot yet resolve the large motions of dimer dissociation and recovery computationally, not least because a structure of the necessary target desensitized dimer state is not yet available for kainate receptors.

For residues that form the cation binding site, molecular dynamics simulations reveal that spatial fluctuations of side chains are quite small when the ion binding sites are fully occupied.[39] These motions increase when either Na or Cl ions are absent from the dimer binding sites. When all binding sites in the dimer assembly are vacant, the random side chain movements are larger still. The dimer is stabilized by intermolecular salt bridges that line the base of the anion site, so the consequences of ion-dependent perturbations are obvious for dimer stability. The consequence of greater disorder is faster dissociation of the LBD dimer, a process that accompanies desensitization. The lid of the anion site undergoes pronounced geometric perturbations, even in the presence of bound chloride. The two salt bridges (between Arg744 and Asp745) that can form here only present in some crystal structures. Instead, one or both of the Arg744 side chains can rotate upwards and open an entry route for Chloride. In the highest resolution dimer structures (such as PDB 3G3F, the GluK2-Glutamate complex),[40] two conformations were detected for these side chains. Dynamic motions of these Arginine residues are also observed in simulations.

Further information about selectivity of the cation site comes from free-energy calculations.[56] The core of the cation site is highly charged and this is responsible for the selection of sodium over larger ions. Lithium, which in electrophysiological experiments strongly resembles sodium, is predicted to bind tightest of all. However, block of desensitization is necessarily a more complex process than the free energy of ion binding, involving conformational changes that are beyond the scope of these computational experiments. These experiments also tentatively suggested a mechanism for cation facilitation of chloride binding. In this model, bound cations partially neutralize the negative electric field at the top of the anion site. In the absence of cations this barrier is high enough to render chloride binding to the intact D1 interface quite unlikely. However, how ions might approach their sites in partly or fully dissociated D1 interface may differ considerably, not least because of the dramatic changes in solvation that must occur when the interface becomes exposed.

Allosteric Modulation of Other Ion Channel Receptors by Monovalent Ions

AMPA-Type Glutamate Receptors

Some reports describe sensitivity of AMPA receptor currents to monovalent ions[46] but no kinetic effects are seen. GluA2 channels seem to be broadly insensitive[39] and GluA1 channels activate almost normally in the absence of all ions,[62] consistent with quite minor ion-dependent changes in activity compared to kainate receptors.

NMDA-Type Glutamate Receptors

Several studies point towards complex allosteric effects mediated by permeant cations in NMDA receptors. External potassium reduced the frequency and the apparent duration of NMDA receptor openings, whilst causing a profound inhibition of the macroscopic current.[63] A separate study suggests that permeant sodium ions reduce the lifetime of the glutamate bound NMDA receptor complex.[64] These studies appear inconsistent, but the methods involved were quite different and so the results cannot be directly compared. Further unusual behaviour was noted in a pore mutant of the NR1 subunit, wherein coupling of permeant ions to channel gating was detected in bi-ionic conditions.[65] The amplitude of the first opening in a group of openings tended to be biased by the internal permeant ion. This suggests that permeant ions can influence gating conformational changes and, perhaps secondarily, ligand affinity. Bound sodium ions were detected in the crystal structure of the ATD of NMDA NR2B. Mutagenesis at these sites, however, failed to demonstrate compelling effects on NMDA receptor currents,[66] possibly because the sites are too low in affinity to be occupied in physiological salt.

GABA and Glycine Receptors

High internal chloride slows the deactivation decay of both native and recombinant glycine receptors.[67] The effect is also clearly detected in single channel currents and can be entirely ascribed to a slowing of the channel shutting rate. Elevating internal chloride above normal baseline levels can also slow GABAergic synaptic currents.[68] Such effects may partly underlie the speeding of inhibitory currents during early postnatal development, as intracellular chloride is reduced by increased expression of cotransporters.[69] Understanding the molecular mechanisms of this kinetic effect is complicated by the voltage sensitivity of synaptic currents, because chloride is the major permeant ion.[70] The quantitative analysis of glycine receptors exceeds what is possible with glutamate receptors, mainly because the single channel currents of glycine receptors are much better characterized.[71] Despite this advantage, the molecular details of chloride modulation are uncertain. The location of the binding site or sites is a completely open question. In the absence of structural information, localizing ion binding sites might be difficult. The voltage-dependence of chloride modulation suggests that, to some extent, either chloride ions within the electric field can stabilize the open state, or ions that are coupled electrostatically or sterically to ions within the field can do so. Eliminating putative chloride binding sites proximal to the electric field is a possible route to localizing the sites, but will by definition hamper functional measurements.

Nicotinic Receptors

Monovalent cations may bind competitively at the acetylcholine binding site in muscle-type nicotinic receptors.[72,73] Larger ions (Cs^+) bind more tightly than Na^+, but half-maximal concentrations for inhibition (~100 mM) are similar to those estimated for the kainate receptor. Of a set of tested mutations, potassium competition was only affected by mutants at position 184 of the epsilon subunit,[74] which either neutralized negative charge (E184Q) or replaced it with a positive charge (E184K). Mutations that altered potassium competition had only very limited effects on ACh dissociation constants. Thus, bound cations may be coupled electrostatically to the charged groups of the ACh molecule, hindering agonist binding at the epsilon-alpha subunit interface *via* electrostatic repulsion..

Trimeric Ion Channel Receptors (ASIC, P2X)

Recombinant P2X7 channels show strong modulation by cations and when Na^+ is replaced, open probability increases, probably because the single channel open time is extended.[75] A definite physiological role for this modulation is seen in the airway cilia, which express P2X receptors.[76] According to these studies, the presence of sodium should curtail channel activation. The structure of the P2X4 receptor did not reveal any sites for monovalent cation binding.[77] However, these crystallization experiments were not designed to detect bound monovalent ions and the resolution of the structure might have been too low to identify sodium ions reliably. Recently, a structurally-related trimeric ligand-gated channel, the acid-sensitive ion channel (ASIC) was shown to bind chloride ions at three symmetrically related sites in the extracellular domain. As in kainate receptors, positively charged residues that coordinate bound chlorides project into subunit interfaces and these residues are conserved across evolution.[78] Positive charges are not conserved in degenerin or ENaC sodium channels, suggesting a specific role for chloride in ASIC function or assembly.

Voltage-Gated Channels

L-type calcium channels are inhibited by reductions in extracellular chloride. Counter-intuitively, this sensitivity appears to arise from molecular features at the intracellular side of the channel.[79] Potassium channels may also be inhibited by intracellular chloride, but only upon disruption of the cytoskeleton.[80] Because intracellular chloride is more likely to vary in normal and disease states than extracellular chloride, these mechanisms offer the opportunity for regulation of channel activity, in addition to crosstalk or coupling to chloride channel activity.

Physiological Situation with Regard to Ions

The question of the physiological role of ion modulation of kainate receptors remains entirely unexplored. There is every possibility that kainate receptors exploit the solemn homeostasis of extracellular sodium and chloride and use these ions as structural features. Certainly, the sodium and chloride binding sites have the right affinity to be modulated by even small changes in the external salt concentration. At least one possibility can be immediately ruled out. Changes in osmotic strength alone do not alter kainate receptor kinetics,[39] so kainate receptors are not osmosensors.

Native Kainate Receptors

Regrettably, there are precious few examples where the kinetics of native kainate receptors have been carefully measured. Cultured hippocampal cells[81] and retinal cells[82,83] are two cases. The desensitization kinetics of receptors in these two cases is similar to that of recombinant receptors composed of the GluK2 subunit. These native channels desensitize profoundly and rapidly and recover very slowly, in contrast to the rapid recovery of AMPA receptors. The kainate receptors expressed by hippocampal cells in culture exhibit similar cation sensitivity to recombinant channels.[48] Thus, it seems highly likely that native channels are also ion-sensitive.

A complication is that the ion sensitivity of recombinant channels has almost exclusively been tested in homomeric receptors. Heteromeric channels incorporating the GluK5 subunit are selectively activated by S-AMPA and these channels demonstrate anion sensitivity similar to homomeric channels in recombinant expression.[31] This is somewhat surprising, because the GluK5 ligand binding core harbors a nonconservative substitution in the anion binding site (Thr748 is instead Asp in GluK5). This variation introduces a negative charge and in a mixed dimer, one might expect that the requirement for chloride might be reduced. However, charge from neighbouring amino acids may compensate in an unforeseen fashion. Such subtleties may be elucidated by heterodimer crystal structures of the kainate receptor ligand binding cores, if they can be obtained.

Since the majority of experiments to assess ion-dependent gating were undertaken, the kainate receptor family has adopted an auxiliary subunit called NETO2.[84] Coexpression of NETO2 with GluK2 slows desensitization and deactivation kinetics and speeds recovery. Are kainate receptors in complex with accessory subunits still as sensitive to ions? This question is completely open, but the recently published GluA2 crystal structure and studies of AMPA receptor-TARP complexes provide some insight to the relative positioning of auxiliary subunits in native complexes. The comparatively late assembly (in the biosynthetic pathway) and rapid dissociation of AMPA-TARP complexes and the substoichiometric ratio of auxiliary proteins to GluR subunits in native complexes[85,86] are consistent with transient associations between GluRs and accessory proteins that occur at the periphery of the channel. In this case, the dimer arrangement of ligand binding domains (and thus the ion binding sites) would be expected to remain largely intact in receptors that complex with NETO2.

Kainate Receptor Trafficking

Have kainate receptors evolved the use of ions as a charge and steric balance in their LBD dimer interfaces, simply because they are so reliably supplied in all brain regions? Such a contention applies only to mature kainate receptors in the cell membrane. During their biosynthesis, the extracellular domains of glutamate receptors are exposed to the ER lumen, which contains high calcium and presumably reduced sodium and chloride levels compared to the extracellular space. Mutations in glutamate receptors that disrupt desensitization, or that weaken ligand binding, have deleterious effects on receptor trafficking both for AMPA receptors[87,88] and kainate receptors.[42,89,90] It is conceivable that dimer dissociation, under the action of ambient glutamate in the ER lumen, is important for receptor export. In this way, low NaCl in the ER would facilitate LBD dimer dissociation for receptor assembly. The desensitization of kainate receptors is fairly profound, independent of the ionic conditions that the receptor is exposed to. Therefore, such a process can only be significant if complete inactivity of kainate receptors is essential or advantageous for ER export.

Coupling of Repetitive Activity to Kainate Receptor Availability

Following repetitive stimulation, sodium influx depletes sodium in the extracellular space. This effect might be particularly pronounced in tight spaces such as the synaptic cleft where diffusion of charged molecules is influenced by electrostatics.[91] This effect is transient but occurs because the NaK-ATPase pumps out sodium quite slowly compared to the rate of channel-mediated influx. The lag of the ATPase creates a window, during which potassium, which has a lower potency for stabilizing the kainate receptor dimer interface, replaces a proportion of the sodium usually found at rest in the extracellular space. Chloride is essentially unchanged. Paternain and colleagues suggested external sodium reductions during repetitive neuronal activity as a physiological role for ion dependent gating of kainate receptors.[48]

In reduced sodium (and increased potassium) the allosteric cation sites of a given kainate receptor have an increased likelihood to become vacant (Fig. 7). In this case, we would expect that chloride site could also become vacant due to the coupled reduction in affinity.[39] Because the dimer interface would be destabilized, binding of glutamate during this window would cause immediate desensitization, rather than activation. Because the recovery of recombinant and native kainate receptors from desensitization is very slow (recovery time constants between 1 and 5 s),[45,81,83] this ion-dependent increase in desensitization might persist and far outlast the recovery of the Na level. Thus a kind of coincidence detection could operate, generating a comparatively long lasting record of repetitive activity through the damping of kainate

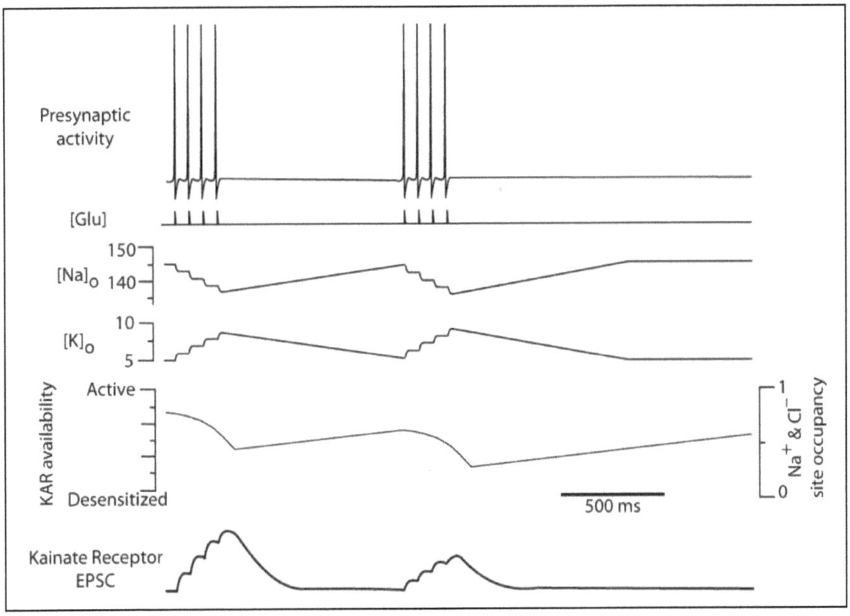

Figure 7. Hypothetical sensing of small fluctuations in external cation concentrations by Kainate receptors. Repetitive presynaptic spiking drives release of glutamate and is accompanied by small shifts in external ion concentrations. These shifts could be accentuated by compact geometry in synaptic and perisynaptic regions. Reduction of external sodium ($[Na]_o$) and increase of external potassium ($[K]_o$) preferentially desensitizes kainate receptors. Slow recovery of kainate receptors from dimer dissociated desensitized states damps subsequent activity. Thus, the kainate receptor EPSC could encode a long-lasting record of strong activity, which could far outlast the rapid re-equilibration of the external ions. Concentrations and timescale are purely illustrative.

receptor activity. Only at a fraction of receptors would be subject to prolonged ion-dependent desensitization, but such tuning could be especially important in downstream modulation of neurotransmitter release, a clear role for kainate receptors.[92,93] Kainate receptors at the Mossy Fiber terminal modulate spike timing precision[94] and this is another role where small changes in charge transfer driven by ion sensitive desensitization could contribute to information processing on a longer temporal scale. But it remains to be demonstrated that such an idea occurs in practice.

GluRs and Salt Concentration Signaling

Although sodium and chloride levels in the cerebral spinal fluid are under tight temporal control in the CNS, sensing of salt concentrations in plasma does occur in the brain. The circumventricular organs are exposed to serum and metabolites because in this part of the brain, the blood-brain barrier is perforated. Thus kainate receptors in the hypothalamus and subfornical organ could be exposed to larger fluctuations in salt concentrations than those in the rest of the brain. A number of ion channels and transporters expressed here participate in salt sensing and control of blood pressure through hormone release.[95,96] NaX channels are expressed in the subfornical organ and are activated by sodium,[97,98] but are found exclusively in glia and not in neurons.[99] Interestingly, NaX deficient mice retain the ability to release vasopressin upon dehydration.[100] In explants from the hypothalamo-neurohypophyseal system, the action of selective antagonists implicates NMDA receptors in this vasopressin release.[101] AMPA and kainate receptors might also be involved,[102] although effects on hormone release are more pronounced with an AMPA selective positive modulator (cyclothiazide) than for a blocker of kainate receptor desensitization (Concanavalin A).

Pharmacological studies also provide evidence that non NMDA receptors are activated in neurons of the subfornical organ leading to release of hormones that control sodium appetite.[103] Kainate receptors expressed in these brain regions have been implicated in hormone secretion which could relate to hypertension,[104] but stimulation with SYM 2081, a selective agonist, did not lead to hormone secretion in explants from hypothalamus and neighbouring brain regions.[105] As in other parts of the brain, glutamate is a major neurotransmitter in this region and thus it is hard to discriminate pharmacological effects on cell-to-cell communication from more specialized signaling roles. The recent advent of selective kainate receptor antagonists may enable a re-evaluation of any possible role for kainate receptors in the central sensing or control of circulatory sodium.

GluRs and Taste Transduction

Glutamate is one of a group of molecules that are sensed as 'Umami'. Although early experiments suggested that metabotropic glutamate receptors are the main transducers of this taste modality,[106] the functional repertoire of Umami sensors is probably far wider.[107] Functional kainate receptors are expressed in taste buds, as revealed by cobalt staining that is stimulated by the selective kainate receptor agonist SYM 2081.[108] Immunostaining shows expression of both GluK2 and GluK4 subunits, but not AMPA receptors. Some taste buds also express NMDA receptors[109] and other data suggests expression of the GluK1 subunit is selectively expressed in Type II and Type II taste cells. The GluK5 subunit is also detected in lingual tissue,[106] suggesting a specific and targeted use of a repertoire of kainate receptors in the tongue.

The cellular localization of kainate receptors in taste buds does not suggest a direct sensing role for dietary glutamate. An intriguing prospect, currently unexplored, is that kainate receptors in taste buds might also be exposed to transients of high dietary salt. Sodium and chloride can penetrate the taste bud through cell-cell apical junctions and this would enable both ions to approach kainate receptors at basally-located signaling sites. The threshold for detection of salt taste is in the range 5 mM, which is consistent with the threshold for NaCl binding and modulation of kainate receptors. Kainate receptors mediating cell-to-cell communication may be involved, if not in detection of salt, then perhaps in modulation of other taste modalities by salt.[110] Several channels have been proposed to sense salt but all have fallen short (reviewed by

Roper, ref. 111). TRPV1 knockout mice retain the ability to taste salt and amiloride, which inhibits the ENaC channel, is ineffective in modulating human salt taste, despite altering salt perception in mice.[112]

Conclusion

A battery of biophysical and biochemical data show that the unusual sensitivity of kainate receptors to monovalent ions derives from specific binding sites which are closely apposed in a key functional intersubunit interface. These sites are exposed to the extracellular space, well away from the membrane pore and are selective for sodium and chloride. The absence of these ions, either because the sites are not saturated by physiological salt, or because ions have been removed, renders kainate receptors nonfunctional. The physiological implications of this strong control of channel activity by ions, which are themselves under strong homeostatic regulation, remain unclear and may be limited to a purely physicochemical adaptation of intersubunit interactions. Given the apparent critical importance of glutamate receptor desensitization in vivo,[113] the strong desensitization of kainate receptors, facilitated by the weak binding of ions in this critical interface, might be more important than previously anticipated.

Acknowledgements

I thank Dr Michael Ashby (NINDS, Bethesda, USA) for a critical reading of the manuscript. Figures were prepared with PYMOL, Google Sketchup, Kaleidagraph and LIGPLOT. Research in the author's laboratory is supported by the Leibniz Association and the Deutsche Forschungsgemeinschaft (EXC 257).

References

1. Miller DJ. Sydney Ringer; physiological saline, calcium and the contraction of the heart. J Physiol 2004; 555:585-587.
2. Jordan J, Shannon JR, Grogan E et al. A potent pressor response elicited by drinking water. Lancet 1999; 353:723.
3. Wells CM, Di Cera E. Thrombin is a Na(+)-activated enzyme. Biochemistry 1992; 31:11721-11730.
4. Perutz MF, Shih DT, Williamson D. The chloride effect in human haemoglobin. A new kind of allosteric mechanism. J Mol Biol 1994; 239:555-560.
5. Hille B. Ion Channels of Excitable Membranes, 3rd Edition. Sunderland, MA, USA: Sinauer Associates; 2001.
6. Eisenman G. Cation selective glass electrodes and their mode of operation. Biophys J 1962; 2:259-323.
7. Nagem RA, Dauter Z, Polikarpov I. Protein crystal structure solution by fast incorporation of negatively and positively charged anomalous scatterers. Acta Crystallogr D Biol Crystallogr 2001; 57:996-1002.
8. Harding MM. Small revisions to predicted distances around metal sites in proteins. Acta Crystallogr D Biol Crystallogr 2006; 62:678-682.
9. Noskov SY, Roux B. Control of ion selectivity in LeuT: two Na^+ binding sites with two different mechanisms. J Mol Biol 2008; 377:804-818.
10. Müller P, Köpke S, Sheldrick GM. Is the bond-valence method able to identify metal atoms in protein structures? Acta Crystallogr D Biol Crystallogr 2003; 59:32-37.
11. Nayal M, Cera ED. Valence screening of water in protein crystals reveals potential Na^+ binding sites. J Mol Biol 1996; 256:228-234.
12. Pidcock E, Moore GR. Structural characteristics of protein binding sites for calcium and lanthanide ions. J Biol Inorg Chem2001; 6:479-489.
13. Erreger K, Chen PE, Wyllie DJ et al. Glutamate receptor gating. Crit Rev Neurobiol 2004; 16:187-224.
14. Mayer ML. Glutamate receptor ion channels. Curr Opin Neurobiol 2005; 15:282-288.
15. Sobolevsky AI, Rosconi MP, Gouaux E. X-ray structure, symmetry and mechanism of an AMPA-subtype glutamate receptor. Nature 2009; 462:745-756.
16. Das U, Kumar J, Mayer ML et al. Domain organization and function in GluK2 subtype kainate receptors. Proc Natl Acad Sci USA 2010:-.
17. Rosenmund C, Stern-Bach Y, Stevens CF. The tetrameric structure of a glutamate receptor channel. Science 1998; 280:1596-1599.
18. Jin R, Banke TG, Mayer ML et al. Structural basis for partial agonist action at ionotropic glutamate receptors. Nat Neurosci 2003; 6:803-810.

19. Verdoorn TA, Burnashev N, Monyer H et al. Structural determinants of ion flow through recombinant glutamate receptor channels. Science 1991; 252:1715-1718.
20. Burnashev N, Zhou Z, Neher E et al. Fractional calcium currents through recombinant GluR channels of the NMDA, AMPA and kainate receptor subtypes. J Physiol 1995; 485 (Pt 2):403-418.
21. Burnashev N, Villarroel A, Sakmann B. Dimensions and ion selectivity of recombinant AMPA and kainate receptor channels and their dependence on Q/R site residues. J Physiol 1996; 496 (Pt 1):165-173.
22. Köhler M, Burnashev N, Sakmann B et al. Determinants of Ca^{2+} permeability in both TM1 and TM2 of high affinity kainate receptor channels: diversity by RNA editing. Neuron 1993; 10:491-500.
23. Stern-Bach Y, Bettler B, Hartley M et al. Agonist selectivity of glutamate receptors is specified by two domains structurally related to bacterial amino acid-binding proteins. Neuron 1994; 13:1345-1357.
24. Ayalon G, Segev E, Elgavish S et al. Two regions in the N-terminal domain of ionotropic glutamate receptor 3 form the subunit oligomerization interfaces that control subtype-specific receptor assembly. J Biol Chem 2005; 280:15053-15060.
25. Frischknecht R, Heine M, Perrais D et al. Brain extracellular matrix affects AMPA receptor lateral mobility and short-term synaptic plasticity. Nat Neurosci 2009; 12:897-904.
26. Saglietti L, Dequidt C, Kamieniarz K et al. Extracellular interactions between GluR2 and N-cadherin in spine regulation. Neuron 2007; 54:461-477.
27. Kumar J, Schuck P, Jin R et al. The N-terminal domain of GluR6-subtype glutamate receptor ion channels. Nat Struct Mol Biol 2009; 16:631-638.
28. Nanao MH, Green T, Stern-Bach Y et al. Structure of the kainate receptor subunit GluR6 agonist-binding domain complexed with domoic acid. Proceedings of the National Academy of Sciences of the United States of America 2005; 102:1708-1713.
29. Clayton A, Siebold C, Gilbert RJC et al. Crystal Structure of the GluR2 Amino-Terminal Domain Provides Insights into the Architecture and Assembly of Ionotropic Glutamate Receptors. J Mol Biol 2009; 392:1125-1132.
30. Horning MS, Mayer ML. Regulation of AMPA receptor gating by ligand binding core dimers. Neuron 2004; 41:379-388.
31. Plested AJR, Mayer ML. Structure and mechanism of kainate receptor modulation by anions. Neuron 2007; 53:829-841.
32. Gielen M, Le Goff A, Stroebel D et al. Structural rearrangements of NR1/NR2A NMDA receptors during allosteric inhibition. Neuron 2008; 57:80-93.
33. Gielen M, Siegler Retchless B, Mony L et al. Mechanism of differential control of NMDA receptor activity by NR2 subunits. Nature 2009; 459:703-707.
34. Anson LC, Chen PE, Wyllie DJ et al. Identification of amino acid residues of the NR2A subunit that control glutamate potency in recombinant NR1/NR2A NMDA receptors. J Neurosci 1998; 18:581-589.
35. Jin R, Clark S, Weeks AM et al. Mechanism of positive allosteric modulators acting on AMPA receptors. J Neurosci 2005; 25:9027-9036.
36. Sun Y, Olson R, Horning M et al. Mechanism of glutamate receptor desensitization. Nature 2002; 417:245-253.
37. Mayer ML. Crystal structures of the GluR5 and GluR6 ligand binding cores: molecular mechanisms underlying kainate receptor selectivity. Neuron 2005; 45:539-552.
38. Mayer ML, Ghosal A, Dolman NP et al. Crystal structures of the kainate receptor GluR5 ligand binding core dimer with novel GluR5-selective antagonists. J Neurosci 2006; 26:2852-2861.
39. Plested AJR, Vijayan R, Biggin PC et al. Molecular basis of kainate receptor modulation by sodium. Neuron 2008; 58:720-735.
40. Chaudhry C, Weston MC, Schuck P et al. Stability of ligand-binding domain dimer assembly controls kainate receptor desensitization. EMBO J 2009; 28:1518-1530.
41. Chaudhry C, Plested AJ, Schuck P et al. Energetics of glutamate receptor ligand binding domain dimer assembly are modulated by allosteric ions. Proc Natl Acad Sci USA 2009; 106:12329-12334.
42. Fleck MW, Cornell E, Mah SJ. Amino-acid residues involved in glutamate receptor 6 kainate receptor gating and desensitization. J Neurosci 2003; 23:1219-1227.
43. Perouansky M, Grantyn R. Separation of quisqualate- and kainate-selective glutamate receptors in cultured neurons from the rat superior colliculus. J Neurosci 1989; 9:70-80.
44. Egebjerg J, Bettler B, Hermans-Borgmeyer I et al. Cloning of a cDNA for a glutamate receptor subunit activated by kainate but not AMPA. Nature 1991; 351:745-748.
45. Bowie D, Lange GD. Functional stoichiometry of glutamate receptor desensitization. J Neurosci 2002; 22:3392-3403.
46. Bowie D. External anions and cations distinguish between AMPA and kainate receptor gating mechanisms. J Physiol 2002; 539:725.

47. Balannik V, Menniti FS, Paternain AV et al. Molecular mechanism of AMPA receptor noncompetitive antagonism. Neuron 2005; 48:279-288.
48. Paternain AV, Cohen A, Stern-Bach Y et al. A role for extracellular Na^+ in the channel gating of native and recombinant kainate receptors. J Neurosci 2003; 23:8641-8648.
49. Gonzalez J, Du M, Parameshwaran K et al. Role of dimer interface in activation and desensitization in AMPA receptors. Proc Natl Acad Sci USA 2010; 107:9891-9896.
50. Plested AJ, Mayer ML. AMPA receptor ligand binding domain mobility revealed by functional cross linking. J Neurosci 2009; 29:11912-11923.
51. Armstrong N, Jasti J, Beich-Frandsen M et al. Measurement of conformational changes accompanying desensitization in an ionotropic glutamate receptor. Cell 2006; 127:85-97.
52. Weston MC, Schuck P, Ghosal A et al. Conformational restriction blocks glutamate receptor desensitization. Nat Struct Mol Biol 2006; 13:1120-1127.
53. Stern-Bach Y, Russo S, Neuman M et al. A point mutation in the glutamate binding site blocks desensitization of AMPA receptors. Neuron 1998; 21:907-918.
54. Zhang Y, Nayeem N, Nanao MH et al. Interface interactions modulating desensitization of the kainate-selective ionotropic glutamate receptor subunit GluR6. J Neurosci 2006; 26:10033-10042.
55. Nayeem N, Zhang Y, Schweppe DK et al. A nondesensitizing kainate receptor point mutant. Mol Pharmacol 2009; 76:534-542.
56. Vijayan R, Plested AJ, Mayer ML et al. Selectivity and cooperativity of modulatory ions in a neurotransmitter receptor. Biophys J 2009; 96:1751-1760.
57. Wong AY, MacLean DM, Bowie D. Na^+/Cl^- dipole couples agonist binding to kainate receptor activation. J Neurosci 2007; 27:6800-6809.
58. Plested AJ, Mayer ML. Engineering a high-affinity allosteric binding site for divalent cations in kainate receptors. Neuropharmacology 2009; 56:114-120.
59. Wollmuth LP, Kuner T, Jatzke C et al. The Lurcher mutation identifies delta 2 as an AMPA/kainate receptor-like channel that is potentiated by $Ca(^{2+})$. J Neurosci 2000; 20:5973-5980.
60. Hansen KB, Naur P, Kurtkaya NL et al. Modulation of the dimer interface at ionotropic glutamate-like receptor delta} 2 by D-serine and extracellular calcium. Journal of Neuroscience 2009; 29:907.
61. Koshland Jr DE. Application of a theory of enzyme specificity to protein synthesis. Proc Natl Acad Sci USA 1958; 44:98.
62. Wong AY, Fay AM, Bowie D. External ions are coactivators of kainate receptors. J Neurosci 2006; 26:5750-5755.
63. Ozawa S, Iino M, Tsuzuki K. Suppression by extracellular K^+ of N-methyl-D-aspartate responses in cultured rat hippocampal neurons. J Neurophysiol 1990; 64:1361-1367.
64. Nahum-Levy R, Tam E, Shavit S et al. Glutamate but not glycine agonist affinity for NMDA receptors is influenced by small cations. J Neurosci 2002; 22:2550-2560.
65. Schneggenburger R, Ascher P. Coupling of permeation and gating in an NMDA-channel pore mutant. Neuron 1997; 18:167-177.
66. Karakas E, Simorowski N, Furukawa H. Structure of the zinc-bound amino-terminal domain of the NMDA receptor NR2B subunit. EMBO J 2009; 28:3910-3920.
67. Pitt SJ, Sivilotti LG, Beato M. High intracellular chloride slows the decay of glycinergic currents. J Neurosci 2008; 28:11454-11467.
68. Houston CM, Bright DP, Sivilotti LG et al. Intracellular chloride ions regulate the time course of GABA-mediated inhibitory synaptic transmission. J Neurosci 2009; 29:10416-10423.
69. Rivera C, Voipio J, Payne JA et al. The K^+/Cl^- cotransporter KCC2 renders GABA hyperpolarizing during neuronal maturation. Nature 1999; 397:251-255.
70. Gill SB, Veruki ML, Hartveit E. Functional properties of spontaneous IPSCs and glycine receptors in rod amacrine (AII) cells in the rat retina. J Physiol 2006; 575:739-759.
71. Burzomato V, Beato M, Groot-Kormelink PJ et al. Single-channel behavior of heteromeric alpha1beta glycine receptors: an attempt to detect a conformational change before the channel opens. J Neurosci 2004; 24:10924-10940.
72. Manthey AA. Delay of desensitization onset by potassium ion in voltage-clamped frog muscle fibers. Am J Physiol Cell Physiol 1985; 249:C435-C446.
73. Akk G, Auerbach A. Inorganic, monovalent cations compete with agonists for the transmitter binding site of nicotinic acetylcholine receptors. Biophys J 1996; 70:2652-2658.
74. Akk G, Zhou M, Auerbach A. A mutational analysis of the acetylcholine receptor channel transmitter binding site. Biophysical Journal 1999; 76:207-218.
75. Riedel T, Schmalzing G, Markwardt F. Influence of extracellular monovalent cations on pore and gating properties of P2X7 receptor-operated single-channel currents. Biophys J 2007; 93:846-858.
76. Ma W, Korngreen A, Uzlaner N et al. Extracellular sodium regulates airway ciliary motility by inhibiting a P2X receptor. Nature 1999; 400:894-897.

77. Kawate T, Michel JC, Birdsong WT et al. Crystal structure of the ATP-gated P2X(4) ion channel in the closed state. Nature 2009; 460:592-598.
78. Jasti J, Furukawa H, Gonzales EB et al. Structure of acid-sensing ion channel 1 at 1.9 A resolution and low pH. Nature 2007; 449:316-323.
79. Babai N, Kanevsky N, Dascal N et al. Anion-sensitive regions of L-type CaV1.2 calcium channels expressed in HEK293 cells. PLoS One 2010; 5:e8602.
80. Bekar LK, Loewen ME, Forsyth GW et al. Chloride concentration affects Kv channel voltage-gating kinetics: Importance of experimental anion concentrations. Brain Res Bull 2005; 67:142-146.
81. Paternain AV, Rodríguez-Moreno A, Villarroel A et al. Activation and desensitization properties of native and recombinant kainate receptors. Neuropharmacology 1998; 37:1249-1259.
82. DeVries SH, Schwartz EA. Kainate receptors mediate synaptic transmission between cones and 'Off' bipolar cells in a mammalian retina. Nature 1999; 397:157-160.
83. DeVries SH. Bipolar cells use kainate and AMPA receptors to filter visual information into separate channels. Neuron 2000; 28:847-856.
84. Zhang W, St-Gelais F, Grabner CP et al. A transmembrane accessory subunit that modulates kainate-type glutamate receptors. Neuron 2009; 61:385-396.
85. Kim KS, Yan D, Tomita S. Assembly and stoichiometry of the AMPA receptor and transmembrane AMPA receptor regulatory protein complex. J Neurosci 2010; 30:1064-1072.
86. Morimoto-Tomita M, Zhang W, Straub C et al. Autoinactivation of neuronal AMPA receptors via glutamate-regulated TARP interaction. Neuron 2009; 61:101-112.
87. Penn AC, Williams SR, Greger IH. Gating motions underlie AMPA receptor secretion from the endoplasmic reticulum. EMBO J 2008; 27:3056-3068.
88. Greger IH, Akamine P, Khatri L et al. Developmentally regulated, combinatorial RNA processing modulates AMPA receptor biogenesis. Neuron 2006; 51:85-97.
89. Mah SJ, Cornell E, Mitchell NA et al. Glutamate receptor trafficking: endoplasmic reticulum quality control involves ligand binding and receptor function. J Neurosci 2005; 25:2215-2225.
90. Valluru L, Xu J, Zhu Y et al. Ligand binding is a critical requirement for plasma membrane expression of heteromeric kainate receptors. J Biol Chem 2005; 280:6085-6093.
91. Sylantyev S, Savtchenko LP, Niu YP et al. Electric fields due to synaptic currents sharpen excitatory transmission. Science (New York, N.Y.) 2008; 319:1845-1849.
92. Breustedt J, Schmitz D. Assessing the role of GLUK5 and GLUK6 at hippocampal mossy fiber synapses. J Neurosci 2004; 24:10093-10098.
93. Chittajallu R, Vignes M, Dev KK et al. Regulation of glutamate release by presynaptic kainate receptors in the hippocampus. Nature 1996; 379:78-81.
94. Sachidhanandam S, Blanchet C, Jeantet Y et al. Kainate receptors act as conditional amplifiers of spike transmission at hippocampal mossy fiber synapses. J Neurosci 2009; 29:5000-5008.
95. Orlov SN, Mongin AA. Salt-sensing mechanisms in blood pressure regulation and hypertension. Am J Physiol Heart Circ Physiol 2007; 293:H2039-H2053.
96. de Wardener HE. The hypothalamus and hypertension. Physiol Rev2001; 81:1599-1658.
97. Hiyama TY, Watanabe E, Ono K et al. Na(x) channel involved in CNS sodium-level sensing. Nat Neurosci 2002; 5:511-512.
98. Hiyama TY, Watanabe E, Okado H et al. The subfornical organ is the primary locus of sodium-level sensing by Na(x) sodium channels for the control of salt-intake behavior. J Neurosci 2004; 24:9276-9281.
99. Shimizu H, Watanabe E, Hiyama TY et al. Glial Nax channels control lactate signaling to neurons for brain [Na^+] sensing. Neuron 2007; 54:59-72.
100. Nagakura A, Hiyama TY, Noda M. Na(x)-deficient mice show normal vasopressin response to dehydration. Neurosci Lett 2010; 472:161-165.
101. Swenson KL, Badre SE, Morsette DJ et al. N-methyl-D-aspartic acid stimulation of vasopressin release: role in osmotic regulation and modulation by gonadal steroids. J Neuroendocrinol 1998; 10:679-685.
102. Sladek CD, Badre SE, Morsette DJ et al. Role of nonNMDA receptors in osmotic and glutamate stimulation of vasopressin release: effect of rapid receptor desensitization. J Neuroendocrinol 1998; 10:897-903.
103. Xu SH, Inenaga K, Honda E et al. Glutamatergic synaptic inputs activate neurons in the subfornical organ through nonNMDA receptors. J Auton Nerv Syst 2000; 78:177-180.
104. Morsette DJ, Swenson KL, Badre SE et al. Regulation of vasopressin release by ionotropic glutamate receptor agonists. Adv Exp Med Biol 1998; 449:129-130.
105. Morsette DJ, Sidorowicz H, Sladek CD. Role of nonNMDA receptors in vasopressin and oxytocin release from rat hypothalamo-neurohypophysial explants. Am J Physiol Regul Integr Comp Physiol 2001; 280:R313-R322.

106. Chaudhari N, Yang H, Lamp C et al. The taste of monosodium glutamate: membrane receptors in taste buds. J Neurosci 1996; 16:3817-3826.
107. Chaudhari N, Pereira E, Roper SD. Taste receptors for umami: the case for multiple receptors. Am J Clin Nutr 2009; 90:738S-742S.
108. Chung KM, Lee SB, Heur R et al. Glutamate-induced cobalt uptake elicited by kainate receptors in rat taste bud cells. Chem Senses 2005; 30:137-143.
109. Caicedo A, Jafri MS, Roper SD. In situ Ca^{2+} imaging reveals neurotransmitter receptors for glutamate in taste receptor cells. J Neurosci 2000; 20:7978-7985.
110. O'Mahony M, Kingsley L, Harji A et al. What sensation signals the salt taste threshold? Chem. Senses 1976; 2:177-188.
111. Roper SD. Signal transduction and information processing in mammalian taste buds. Pflugers Arch 2007; 454:759-776.
112. Chandrashekar J, Kuhn C, Oka Y et al. The cells and peripheral representation of sodium taste in mice. Nature 2010; 464:297-301.
113. Christie LA, Russell TA, Xu J et al. Contractor A. AMPA receptor desensitization mutation results in severe developmental phenotypes and early postnatal lethality. Proc Natl Acad Sci USA 2010; 107:9412-9417.

Chapter 10

BTB-Kelch Proteins and Ubiquitination of Kainate Receptors

John Marshall,* Leslie A.C. Blair and Jeffrey D. Singer

Abstract

Kainate receptors (KAR) form a class of glutamate receptors that have been implicated in epilepsy, stroke, Alzheimer's and neuropathic pain.[1] KAR subtypes are known to be segregated to specific locations within neurons and play significant roles in synaptic transmission and plasticity.[2] Increasing evidence suggests a the role for ubiqutination in regulating the number of synaptic neurotransmitter receptors.[3-5] The ubiquitin pathway consists of activation (E1), conjugation (E2) and ligation (E3). Cullins form the largest family of E3 ligase complexes. We have recently shown that the BTB/Kelch domain proteins, actinfilin and mayven, bind both Cul3 and specific KAR subtypes (GluR6 and GluR5-2b) to target these KARs for ubiquitination and degradation.[5] In this chapter we will review how these interactions occur, what they mean for the stability of KARs and their associated proteins and how, in turn, they may affect synaptic functions in the central nervous system.

Introduction

KARs are found pre and postsynaptically and have been implicated in the etiology of epilepsy, as well as stroke-induced neurodegeneration and Huntington's disease.[6,7] Epilepsy occurs when inhibitory adaptation is unable to prevent excess neural activity. The developing cortex is particularly vulnerable and a number of syndromes are associated with epilepsy at an early age.[8] Postsynaptic injection of kainic acid causes epileptiform discharges and the death of hippocampal CA3 pyramidal neurons.[9,10] Moreover, KARs are subject to developmental and activity-dependent regulation at thalamocortical synapses and are likely to play an important role in the development of hippocampal synaptic circuits.[11-13] KARs act as excitatory glutamate-gated ion channels: KAR-mediated excitatory postsynaptic currents were first described at mossy fiber-CA3 pyramidal cell synapses,[14,15] while presynaptically, activation of KARs on inhibitory interneurons decreases GABA release which acts to enhance electrical activity, suggesting that presynaptic KARs may be epileptigeneic.[16] Notably, the GluR6 subtype of KARs can also increase neuronal excitability via metabotropic regulation of potassium channels.[17] To treat pathological conditions it will be crucial to understand the molecular mechanisms that determine localization of specific KARs to specific membrane domains.

Within KARs, there is a considerable diversity of properties, including unitary channel conductance, Ca^{2+} permeability and rectification, which arise from differences in receptor subunit composition and RNA editing of GluR5 and GluR6.[18-20] KARs are tetramers that can be assembled from any one of five receptor subunits encoded by two separate gene families.

*Corresponding Author: John Marshall—Department of Molecular Pharmacology, Physiology and Biotechnology, Brown University, Providence, Rhode Island 02912, USA.
Email: John_Marshall@brown.edu

Kainate Receptors: Novel Signaling Insights, edited by Antonio Rodríguez-Moreno and Talvinder S. Sihra.

The first of these includes receptor subunits GluR5, -6 and -7. Each of these subunits can form functional homomeric ion channels or heteromeric mixtures that appear to assemble promiscuously with any available subunit GluR5, -6 or -7.[21-23] Alternative splicing of GluR5 yields two isoforms:[21] GluR5-1 and GluR5-2, which has three additional splice variants possessing distinct C-terminal sequences. The shortest variant is designated GluR5-2a, while additional exons located in the C-terminal region give rise to GluR5-2b and GluR5-2c; these variants share a C-terminal type 1 PDZ-binding domain that is absent in GluR5-2a.[18] The second gene family consists of KA1 and KA2 subunits that are functional only when expressed as heteromeric assemblies with GluR5, -6 or -7.[23,24]

Alternative splicing and RNA editing of ionotropic glutamate receptors play important roles in receptor assembly and trafficking.[25-27] Regulatory steps in the assembly of KA2-containing KARs are governed by at least two trafficking signals located in the cytoplasmic terminal (C-tail) of the KA2 subunit. The first is an arginine-rich motif which operates as an endoplasmic reticulum (ER) retention signal preventing the insertion of homomeric KA2 receptors into the plasma membrane.[28] The second is a di-leucine motif which mediates the internalization and subsequent relocalization of surface-expressed KA2 subunits.[28] Similarly, GluR5-2b carries a positively charged amino acid motif that acts as a novel ER retention signal.[29] In contrast, GluR6, which is highly expressed at the plasma membrane, has a forward trafficking signal in its C-terminal domain critical for ER exit.[30,31]

These differences in targeting appear to convey specific roles to specific KAR subtypes: In GluR6 knockout mice, mossy fiber long-term potentiation (LTP) was reduced, whereas GluR5 knockout mice exhibited normal LTP.[32] The activation of KARs also modulates neurotransmitter release from a number of hippocampal synapses, including GABA release at inhibitory terminals that synapse onto CA1 pyramidal cells.[16,33-35] In hippocampal slices, kainate depresses GABA-mediated synaptic inhibition and increases the firing rate of interneurons. These effects are explained by two populations of KARs in CA1 interneurons: GluR6/KA2 located in the somatodendritic compartment and GluR5-GluR6 or GluR5-KA2 at presynaptic terminals.[35] It is anticipated that this segregation of KARs will allow us to design drugs that specifically target each function.

Recent evidence suggests that the ubiquitin-proteasome pathway and synaptic activity affects the composition of postsynaptic proteins.[36-38] The addition of ubiquitin to proteins leads to a variety of fates for the tagged proteins, one of which is degradation by the 26S proteasome.[39] A family of proteins called E3 ligases determines the specificity of ubiquitin addition. E3 ligases frequently consist of a complex of proteins that act together for specific substrate binding and ubiquitin ligation activity. Two major families of E3 ligases have been described: the HECT-domain family that is defined by its homology to the C-terminus of E6-associated protein (E6AP) and the RING family that contains either an intrinsic RING-finger domain or an associated RING-finger protein subunit essential for ubiquitin ligase activity.[40] One of the best-characterized subset of the RING E3 ligases is the Skp1/Cul1/F-box protein complex (SCF), in which Cul1 binds an adaptor molecule, Skp1.[41,42] Skp1 associates with an F-box protein that in turn binds a phosphorylated substrate. The Cul1 component of the SCF E3 ligase belongs to an evolutionarily conserved family of proteins known as cullins, of which there are six closely related members (Cul1, 2, 3, 4A, 4B and 5) and three distant relatives (Cul7, Parc and APC2).

A major class of Kelch proteins, defined as containing a 6-fold tandem "kelch" element,[43] contains an N-terminal BTB/POZ domain and C-terminal kelch repeats and targets different substrates to the Cul3-Roc1 catalytic core.[44] For example, the BTB-Kelch protein Keap1, a negative regulator of the transcription factor Nrf2, binds Cul3 and Nrf2 via its BTB and kelch domains, respectively, targeting Nrf2 for ubiquitination and subsequent degradation by the proteasome.[45] The BTB-Kelch family also includes the closely related protein mayven, an actin-binding protein and gigaxonin, which is mutated in a human autosomal recessive neurodegenerative disorder named giant axonal neuropathy.[46] Mutations in E3 ubiquitin ligases have also been associated with Parkinson's disease and breast cancer.[47]

KAR Regulation by the Ubiquitin-Proteasome Pathway

To search for proteins involved in the regulation of KARs, we performed a yeast two-hybrid screen of an adult rat brain cDNA library using the C-terminus of GluR6 as bait. Strong interactions were detected between GluR6 and actinfilin. Actinfilin (AF) is a novel BTB/Kelch protein that was identified as a brain-specific actin-binding protein in postsynaptic densities (PSDs).[48] Co-immunoprecipitation studies performed using HEK293 cell and rat brain extracts show that actinfilin binds GluR6 and GluR5-2b, but not with other glutamate receptors and ion channels.[5] Because actinfilin is highly similar to another brain BTB/Kelch protein member, mayven[49] (55% amino acid identity), we cloned this cDNA and, upon expression, found that it also co-immunoprecipitated with GluR6.

Actinfilin was found to interact with Cul3 to promote proteasomal degradation of GluR6 in vitro and in vivo.[5] Expression of GluR6 with an HA-tagged ubiquitin (Ub-HA) in HEK293 cells showed a characteristic ladder indicating that GluR6 was ubiquitinated. Conversely, treatment with the 26S proteasomal inhibitor, MG132, greatly enhanced ubiquitination and stabilized GluR6 expression, demonstrating that GluR6 protein is fairly short-lived. Furthermore, co-immunoprecipitation studies verified that actinfilin interacts with Cul3, but not Cul1, in brain. The interactions of cullins with their adaptors often cause mutual degradation. Importantly, Cul3 appears to specifically regulate KAR levels in vivo (Fig. 1): In synaptosomes prepared from heterozygous Cul3 knockout mice, GluR6 levels are substantially increased, a small effect is observed on KA2 levels, but significantly, no effect on AMPA or NMDA receptors can be detected. These data suggest that Cul3 promotes degradation of KARs.

We also found that actinfilin was localized synaptically in hippocampal and cortical neurons and that it negatively regulates KAR expression (Fig. 2). A high degree of colocalization of AF and GluR6 was observed in dendritic spines. To establish tools to determine the role of actinifilin in the trafficking and/or synaptic localization of GluR6, we have developed a short hairpin inhibitory RNA (RNAi) to actinfilin that eliminates actinifilin expression (Fig. 2D). Specifically, we found that decreasing actinfilin levels via RNAi and overexpressing an inactive Cul3 both induced increased

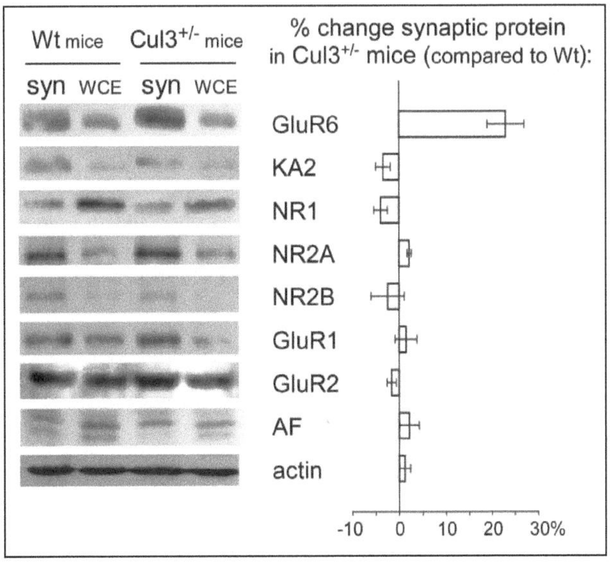

Figure 1. Synaptic GluR6 levels are much higher in synaptosomes (syn) from +/– Cul3 mice than in wild-type (Wt) mice and KA2 show a small increase. Significantly, neither GluR2, NR1 nor NR2B detectably changed. WCE, whole-cell extract.

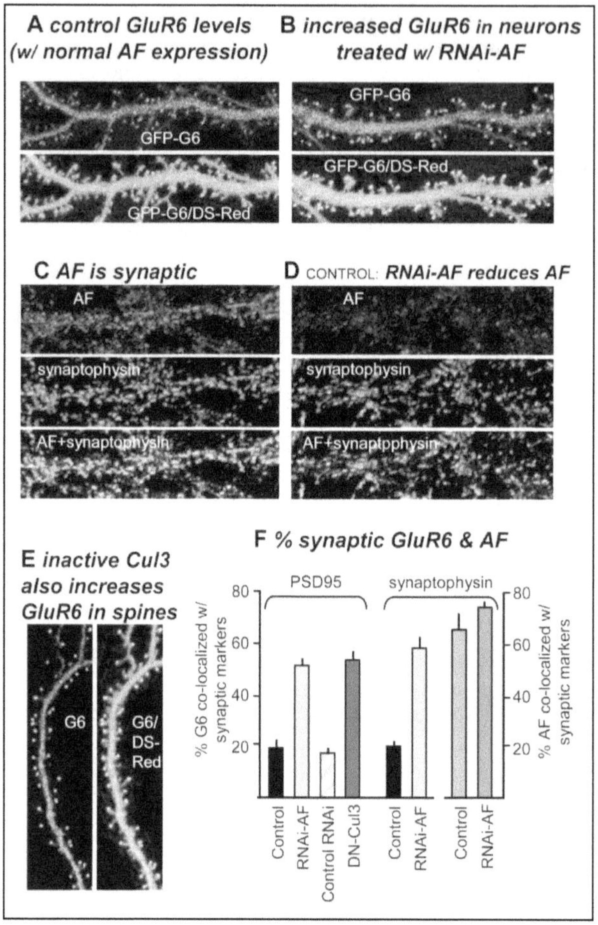

Figure 2. Actinfilin (AF) and Cul3 negatively regulate surface expression of GluR6. A-B) Hippocampal neurons were transfected with GluR6 tagged extracellularly with GFP (GFP-G6) and, to identify cellular morphology, DS-Red. Neurons (A) with normal AF levels exhibit relatively low GluR6 levels, while reducing AF levels with RNAi-AF strongly increases the amount of GluR6 found in dendritic spines (B). C-D) Actinfilin largely colocalizes with synaptic markers (here, synaptophysin), but the RNAi-AF strongly reduces AF levels (see also F). E) Overexpressing a dominant negative Cul3 mutant results in high surface levels of endogenous GluR6. F) Reducing AF increases GluR6 colocalization with synaptic markers: The % colocalization of GluR6 with synaptic markers (PSD-95 or synaptophysin) was determined (means ± SEMs).

surface GluR6 expression at synapses, suggesting that actinfilin-Cul3-mediated degradation may provide an important mechanism for regulating neuronal GluR6.

A model of how actinfilin may link GluR6 to Cul3 and the E3 ubiquitin ligase complex is presented in Figure 3. It should be noted that Cul3 is a component of an E3 ubiquitin ligase complex, also composed of the proteins Nedd8 (N8), Rbx1 and a ubiquitin conjugating enzyme or E2. Actinfilin or mayven would then act as adaptors to link GluR6 or GluR5-2b to this complex, binding the receptors through their kelch domains and Cul3 through their BTB domains and enabling ubiquitination of the receptor.

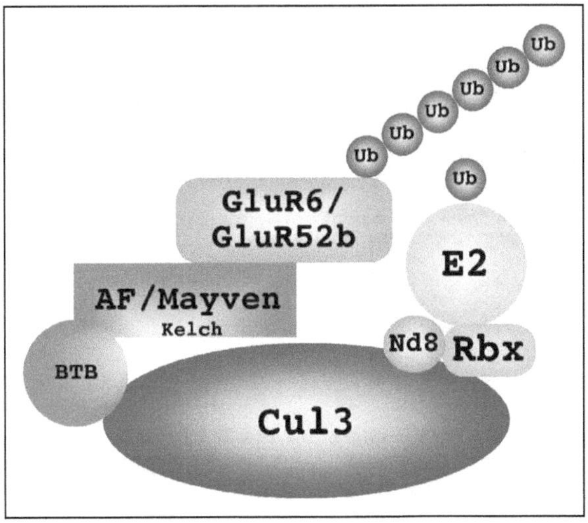

Figure 3. Proposed regulation of KARs by AF and Cul3. The Cul3-based E3 ligase consists of Cul3, which acts as a scaffold to bring several essential proteins in close proximity. These proteins include Nedd 8 (Nd8), Rbx1 and a ubiquitin conjugating enzyme (E2). Substrate specificity is mediated by proteins containing a BTB domain that bind Cul3 in its amino-terminal end. Other domains on the BTB domain-containing proteins are involved in recognition of substrates; in the cases of actinfilin and mayven, these are kelch domains.

AF Regulation of Shank, a Parallel Path to Regulate Surface GluR Expression

AF appears to negatively regulate the formation of dendritic spines in cortical neurons (Fig. 4). Potentially, this effect is not directly via regulating glutamate receptors, but could involve ubiquitination of scaffolding proteins.[50] Shank proteins (Shanks1-3, also known as ProSAPs) constitute a group of postsynaptic, multidomain proteins that link glutamate receptors to intracellular calcium stores[51] and are involved in the enlargement and maturation of dendritic spines.[52] Importantly, Shank is known to be ubiquitinated.[38] Specifically, we found that in synaptosomes prepared from heterozygous Cul3 knockout mice or by decreasing actinfilin levels via RNAi, Shank levels are substantially increased (Fig. 4). Elevating Shank levels by overexpressing Shank3 in cortical neurons increases not only the number of spines, but the likelihood of detecting surface GluR6 in a spine (Fig. 4). These results suggest that actinfilin-Cul3-mediated Shank regulation may provide an important mechanism for regulating spine development and synaptic KAR localization.

Development of Novel Peptidomimetic Drugs for the Treatment of Neurological Disorders

In neurons, actinfilin is preferentially localized to the dendritic spine, a structure rich in actin and critical for synapse formation.[48] Moreover, synaptic activity is known to promote the redistribution of proteasomes from dendritic shafts to spines via an association with actin filaments.[53] Actinfilin and other BTB-Kelch proteins, such as Keap1, have been shown to associate with the actin cytoskeleton and interactions with actin are necessary for Keap1 to regulate Nrf2 levels.[54] Similarly, the actin cytoskeleton and/or other scaffolding proteins may regulate actinfilin function, suggesting possible roles for these proteins in regulating KAR-actinfilin binding or the trafficking of actinfilin-bound GluR6 to the degradation machinery. We find that down-regulating actinfilin in cortical neurons increases both the synaptic localization and the size of synaptic

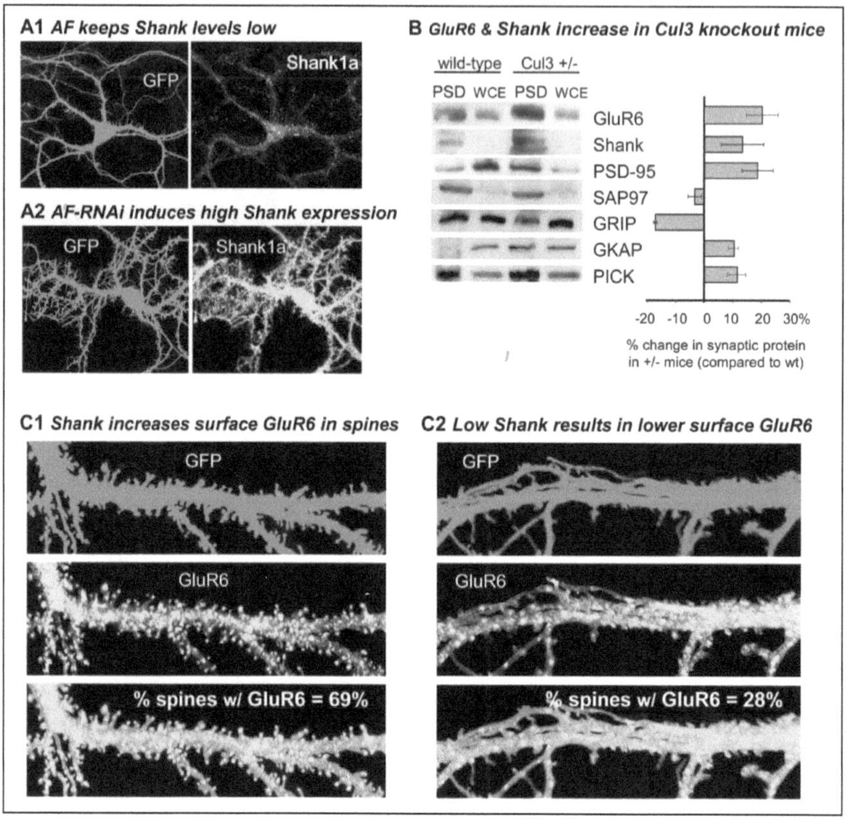

Figure 4. Shank/ProSAP promotes surface GluR6 in dendritic spines and is down-regulated by actinfilin (AF) and Cul3, both of which decrease GluR6. A) AF keeps Shank levels low, but RNAi to AF strongly promotes expression of Shank in hippocampal neurons. B) Reduction of Cul3 ($Cul3^{+/-}$ mice) results in increased GluR6 and Shank in postsynaptic densities (*PSD*). *WCE*, whole cell extract. Also shown is the % change in each synaptic protein in $Cul3^{+/-}$ mice. C) Elevating Shank levels by overexpressing Shank3 in cortical neurons increases not only the number of dendritic spines, but the likelihood of detecting surface GluR6 in a spine (*C1*). For comparison, control neurons with lower Shank levels are also shown (*C2*). The % values indicate the % of spines that have detectable surface GluR6.

GluR6-containing KAR clusters (Fig. 4). Because GluR6 has been implicated in excitotoxic neuronal death, in particular with damage associated with cerebral ischemia, stroke and epileptic seizures, our data imply that actinfilin may provide an important means for ensuring the correct regulation of GluR6 surface expression.

Recently, using NMR to determine the structure of the binding sites regulating GluR interactions with scaffolding and regulatory proteins, we have begun developing compounds that specifically target individual pathways that modulate synaptic GluR levels. One such reagent, CN2180, is a cyclized peptidomimetic compound that targets PSD-95 to inhibit GluR6 clustering.[55] This compound and related analogues, were designed to be membrane permeable, highly selective and resistant to protease digestion. In an in vivo retinal toxicity model,[56] they are taken up rapidly by retinal neurons and attenuate the KA-induced PARP-1 hyperactivation associated with retinal neuron death (Marshall and Goebel, unpublished data). Based on our studies, we

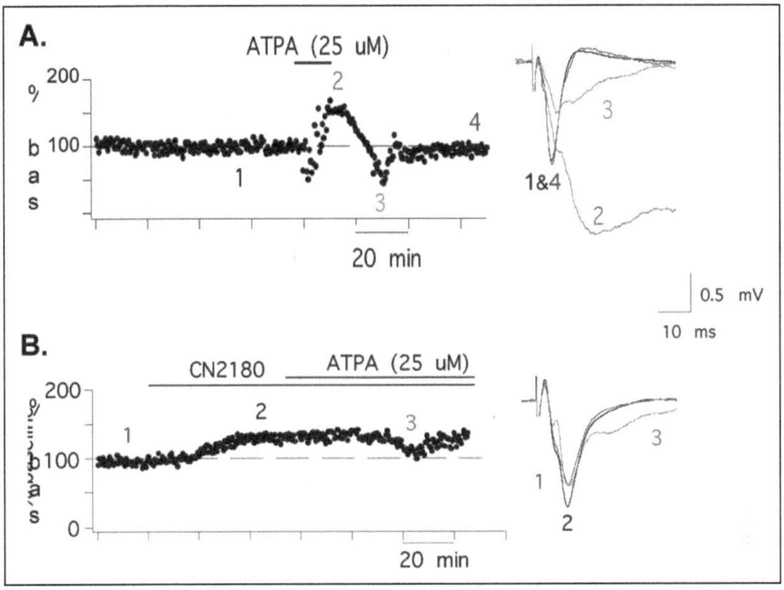

Figure 5. CN2180, a reagent that reduces GluR6 surface expression, strongly reduces seizure-like activity in the motor cortex. A) ATPA, a KAR agonist, induces seizure-like in cortical slices. Time course of field potential amplitudes showing that APTA induces a large, transient increase in field potentials (1,2) followed by loss of evokable potentials (2,3). B) CN2180 pretreatment almost completely blocks the ATPA effects.

conclude that these compounds target the 'main circuit breaker' of the KAR-mediated cell death pathway. The newly developed compounds are expected to improve the tolerance for treatment because they do not affect the ability of glutamate receptors to perform normal neuronal signaling, but they do prevent the disease-related damage.

The effects of these compounds on motor cortical slices are also being examined. ATPA, a GluR5—GluR6/KA2 agonist, causes seizure-like activity, inducing a large, transient increase and widening in field potentials (Fig. 5A). However, we find that CN2180 largely prevents the seizure-like effect (Fig. 5B), without affecting either AMPAR or NMDAR activities. Overall, these studies indicate the applicability of this approach to design drugs specific for KAR-regulation and suggest that actinfilin may also prove a useful therapeutic target to control endogenous synaptic GluR6.

Conclusion

How general is this degradation mechanism for the regulation of kainate receptors? As is shown in Figure 6, there are 61 proteins predicted in the human protein database that contain both a BTB domain and kelch domains. Many of these have been shown to be adaptors for the Cul3-based E3 ligase (red, underlined) in mammals. In addition, KLHL8 which is similar to Kel-8 from *C. elegans* is involved in regulation of the GluR-1 receptor.[57] There are a number of proteins that are very closely related to both mayven and actinfilin which are good candidates for Cul3 substrate adaptors for other related receptors. We think it is very likely that these proteins will serve such a role.

Almost all the other mammalian BTB/Kelch proteins shown to be Cul3 adaptors appear to regulate either receptors or the cytoskeleton. These include KLHL9/KLHL13 which regulate Aurora B kinase,[58] KLHL12 regulates the dopamine D4 receptor[59] and KLHDC5 which regulates

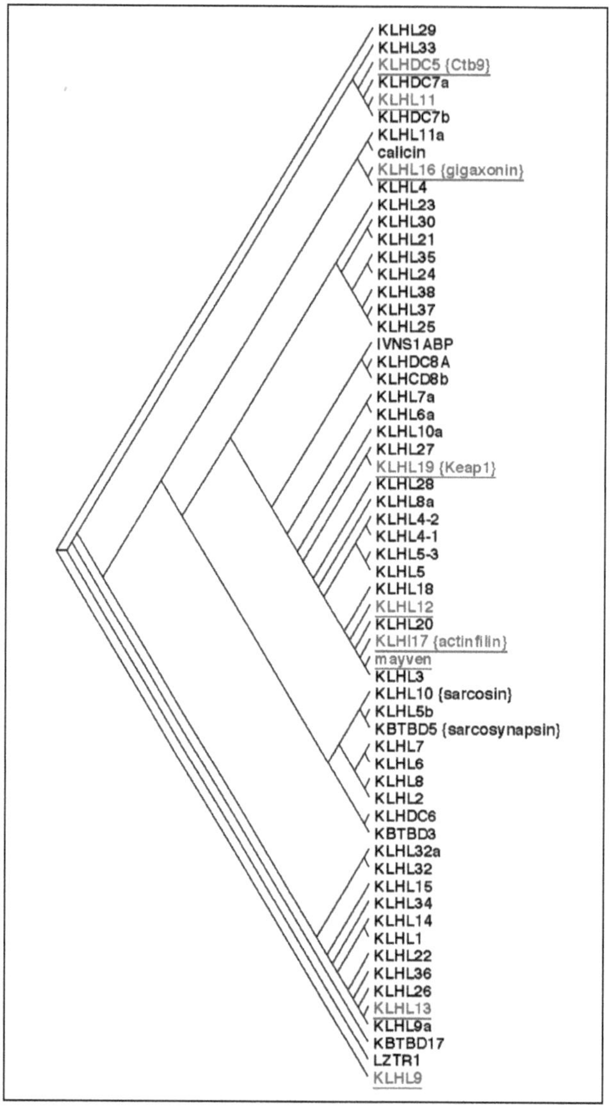

Figure 6. Dendogram showing the 61 human proteins predicted to have both a BTB domain and kelch repeats. The red (underlined) proteins have been shown to act as substrate adaptors for the Cul3-based E3 ligase. The length of the lines connecting the proteins is related to the degree of homology, with shorter distances meaning more closely related (higher percentage of homology). A color version of this figure is available online at www.landesbioscience.com/curie.

KATNA1, a katanin that is involved in microtubule remodeling.[60] Other BTB/Kelch proteins, such as gigaxonin which is involved in peripheral axon development[61] and sarcosin which is involved in cell motility,[62] may also regulate neuronal function through protein ubiquitination. Taken together, Cul3 and its associated BTB/Kelch domain containing substrate adaptors play a critical role in both the development and maintanance of mammalian neurons.

Acknowledgements

We would like to thank Dr. Dennis Goebel (Wayne State University) for conducting the studies on retinal toxicity. We would also like to thank Drs. Greg Salinas and Mengia Rioult-Pedotti for their contributions. This work was supported by NIH R21.2970803 to JM and NIH 1R01GM082940-01A1 to JS.

References

1. Dingledine R, Borges K, Bowie D et al. The glutamate receptor ion channels. Pharmacol Rev 1999; 51:7-61.
2. Lerma J. Roles and rules of kainate receptors in synaptic transmission. Nat Rev Neurosci 2003; 4:481-495.
3. Burbea M, Dreier L, Dittman JS et al. Ubiquitin and AP180 regulate the abundance of GLR-1 glutamate receptors at postsynaptic elements in C. elegans. Neuron 2002; 35:107-120.
4. Colledge M, Snyder EM, Crozier RA et al. Ubiquitination regulates PSD-95 degradation and AMPA receptor surface expression. Neuron 2003; 40:595-607.
5. Salinas GD, Blair LA, Needleman LA et al. Actinfilin is a Cul3 substrate adaptor, linking GluR6 kainate receptor subunits to the ubiquitin-proteasome pathway. J Biol Chem 2006; 281:40164-40173.
6. Ben-Ari Y, Cossart R. Kainate, a double agent that generates seizures: two decades of progress. Trends Neurosci 2000; 23:580-587.
7. Savinainen A, Garcia EP, Dorow D et al. Kainate receptor activation induces mixed lineage kinase-mediated cellular signaling cascades via postsynaptic density protein 95. J Biol Chem 2001; 276:11382-11386.
8. Guerrini R, Andermann E, Avoli M et al. Cortical dysplasias, genetics and epileptogenesis. Adv Neurol 1999; 79:95-121.
9. Lerma J. Kainate reveals its targets. Neuron 1997; 19:1155-1158.
10. Frerking M, Malenka RC, Nicoll RA. Synaptic activation of kainate receptors on hippocampal interneurons. Nat Neurosci 1998; 1:479-486.
11. Kidd FL, Isaac JT. Developmental and activity-dependent regulation of kainate receptors at thalamocortical synapses. Nature 1999; 400:569-573.
12. Lauri SE, Bortolotto ZA, Bleakman D et al. A critical role of a facilitatory presynaptic kainate receptor in mossy fiber LTP. Neuron 2001; 32:697-709.
13. Lauri SE, Segerstrale M, Vesikansa A et al. Endogenous activation of kainate receptors regulates glutamate release and network activity in the developing hippocampus. J Neurosci 2005; 25:4473-4484.
14. Castillo PE, Malenka RC, Nicoll RA. Kainate receptors mediate a slow postsynaptic current in hippocampal CA3 neurons. Nature 1997; 388:182-186.
15. Vignes M, Collingridge GL. The synaptic activation of kainate receptors. Nature 1997; 388:179-182.
16. Rodriguez-Moreno A, Lerma J. Kainate receptor modulation of GABA release involves a metabotropic function. Neuron 1998; 20:1211-1218.
17. Fisahn A, Heinemann SF, McBain CJ. The kainate receptor subunit GluR6 mediates metabotropic regulation of the slow and medium AHP currents in mouse hippocampal neurones. J Physiol 2005; 562(Pt 1):199-203.
18. Sommer B, Burnashev N, Verdoorn TA et al. A glutamate receptor channel with high affinity for domoate and kainate. Embo J 1992; 11:1651-1656.
19. Herb A, Burnashev N, Werner P et al. The KA-2 subunit of excitatory amino acid receptors shows widespread expression in brain and forms ion channels with distantly related subunits. Neuron 1992; 8:775-785.
20. Isaac JT, Mellor J, Hurtado D et al. Kainate receptor trafficking: physiological roles and molecular mechanisms. Pharmacol Ther 2004; 104:163-172.
21. Bettler B, Boulter J, Hermans-Borgmeyer I et al. Cloning of a novel glutamate receptor subunit, GluR5: expression in the nervous system during development. Neuron 1990; 5:583-595.
22. Cui C, Mayer ML. Heteromeric kainate receptors formed by the coassembly of GluR5, GluR6 and GluR7. J Neurosci 1999; 19:8281-8291.
23. Schiffer HH, Swanson GT, Heinemann SF. Rat GluR7 and a carboxy-terminal splice variant, GluR7b, are functional kainate receptor subunits with a low sensitivity to glutamate. Neuron 1997; 19:1141-1146.
24. Wenthold RJ, Trumpy VA, Zhu WS et al. Biochemical and assembly properties of GluR6 and KA2, two members of the kainate receptor family, determined with subunit-specific antibodies. J Biol Chem 1994; 269:1332-1339.
25. Standley S, Roche KW, McCallum J et al. PDZ domain suppression of an ER retention signal in NMDA receptor NR1 splice variants. Neuron 2000; 28:887-898.
26. Scott DB, Blanpied TA, Swanson GT et al. An NMDA receptor ER retention signal regulated by phosphorylation and alternative splicing. J Neurosci 2001; 21:3063-3072.

27. Greger IH, Khatri L, Kong X et al. AMPA receptor tetramerization is mediated by Q/R editing. Neuron 2003; 40:763-774.
28. Ren Z, Riley NJ, Garcia EP et al. Multiple trafficking signals regulate kainate receptor KA2 subunit surface expression. J Neurosci 2003; 23:6608-6616.
29. Ren Z, Riley NJ, Needleman LA et al. Cell surface expression of GluR5 kainate receptors is regulated by an endoplasmic reticulum retention signal. J Biol Chem 2003; 278:52700-52709.
30. Yan S, Sanders JM, Xu J et al. A C-terminal determinant of GluR6 kainate receptor trafficking. J Neurosci 2004; 24:679-691.
31. Jaskolski F, Coussen F, Nagarajan N et al. Subunit composition and alternative splicing regulate membrane delivery of kainate receptors. J Neurosci 2004; 24:2506-2515.
32. Contractor A, Swanson G, Heinemann SF. Kainate receptors are involved in short- and long-term plasticity at mossy fiber synapses in the hippocampus. Neuron 2001; 29:209-216.
33. Clarke VR, Collingridge GL. Characterisation of the effects of ATPA, a GLU(K5) kainate receptor agonist, on GABAergic synaptic transmission in the CA1 region of rat hippocampal slices. Neuropharmacology 2004; 47:363-372.
34. Bureau I, Dieudonne S, Coussen F et al. Kainate receptor-mediated synaptic currents in cerebellar Golgi cells are not shaped by diffusion of glutamate. Proc Natl Acad Sci USA 2000; 97:6838-6843.
35. Christensen JK, Paternain AV, Selak S et al. A mosaic of functional kainate receptors in hippocampal interneurons. J Neurosci 2004; 24:8986-8993.
36. Hegde AN, DiAntonio A. Ubiquitin and the synapse. Nat Rev Neurosci 2002; 3:854-861.
37. Murphey RK, Godenschwege TA. New roles for ubiquitin in the assembly and function of neuronal circuits. Neuron 2002; 36:5-8.
38. Ehlers MD. Activity level controls postsynaptic composition and signaling via the ubiquitin-proteasome system. Nat Neurosci 2003; 6:231-242.
39. Craig KL, Tyers M. The F-box: a new motif for ubiquitin dependent proteolysis in cell cycle regulation and signal transduction. Prog Biophys Mol Biol 1999; 72:299-328.
40. Deshaies RJ. SCF and Cullin/Ring H2-based ubiquitin ligases. Annu Rev Cell Dev Biol 1999; 15:435-467.
41. Furukawa M, Ohta T, Xiong Y. Activation of UBC5 ubiquitin-conjugating enzyme by the RING finger of ROC1 and assembly of active ubiquitin ligases by all cullins. J Biol Chem 2002; 277:15758-15765.
42. Furukawa M, He YJ, Borchers C et al. Targeting of protein ubiquitination by BTB-Cullin 3-Roc1 ubiquitin ligases. Nat Cell Biol 2003; 5:1001-1007.
43. Xue F, Cooley L. kelch encodes a component of intercellular bridges in Drosophila egg chambers. Cell 1993; 72:681-693.
44. Geyer R, Wee S, Anderson S et al. BTB/POZ domain proteins are putative substrate adaptors for cullin 3 ubiquitin ligases. Mol Cell 2003; 12:783-790.
45. Furukawa M, Xiong Y. BTB protein Keap1 targets antioxidant transcription factor Nrf2 for ubiquitination by the Cullin 3-Roc1 ligase. Mol Cell Biol 2005; 25:162-171.
46. Bomont P, Cavalier L, Blondeau F et al. The gene encoding gigaxonin, a new member of the cytoskeletal BTB/kelch repeat family, is mutated in giant axonal neuropathy. Nat Genet 2000; 26:370-374.
47. Yamamoto A, Friedlein A, Imai Y et al. Parkin phosphorylation and modulation of its E3 ubiquitin ligase activity. J Biol Chem 2005; 280:3390-3399.
48. Chen Y, Derin R, Petralia RS et al. Actinfilin, a brain-specific actin-binding protein in postsynaptic density. J Biol Chem 2002; 277:30495-30501.
49. Soltysik-Espanola M, Rogers RA, Jiang S et al. Characterization of Mayven, a novel actin-binding protein predominantly expressed in brain. Mol Biol Cell 1999; 10:2361-2375.
50. Sheng M, Pak DT. Ligand-gated ion channel interactions with cytoskeletal and signaling proteins. Annu Rev Physiol 2000; 62:755-778.
51. Gao L, Blair LA, Salinas GD et al. Insulin-like growth factor-1 modulation of CaV1.3 calcium channels depends on Ca2+ release from IP3-sensitive stores and calcium/calmodulin kinase II phosphorylation of the alpha1 subunit EF hand. J Neurosci 2006; 26:6259-6268.
52. Sala C, Piech V, Wilson NR et al. Regulation of dendritic spine morphology and synaptic function by Shank and Homer. Neuron 2001; 31:115-130.
53. Bingol B, Schuman EM. Activity-dependent dynamics and sequestration of proteasomes in dendritic spines. Nature 2006; 441:1144-1148.
54. Kang MI, Kobayashi A, Wakabayashi N et al. Scaffolding of Keap1 to the actin cytoskeleton controls the function of Nrf2 as key regulator of cytoprotective phase 2 genes. Proc Natl Acad Sci USA 2004; 101:2046-2051.
55. Piserchio A, Salinas GD, Li T et al. Targeting specific PDZ domains of PSD-95; structural basis for enhanced affinity and enzymatic stability of a cyclic peptide. Chem Biol 2004; 11:469-473.

56. Goebel DJ, Winkler BS. Blockade of PARP activity attenuates poly(ADP-ribosyl)ation but offers only partial neuroprotection against NMDA-induced cell death in the rat retina. J Neurochem 2006; 98:1732-1745.
57. Schaefer H, Rongo C. KEL-8 is a substrate receptor for CUL3-dependent ubiquitin ligase that regulates synaptic glutamate receptor turnover. Mol Biol Cell 2006; 17:1250-1260.
58. Sumara I, Peter M. A Cul3-based E3 ligase regulates mitosis and is required to maintain the spindle assembly checkpoint in human cells. Cell Cycle 2007; 6:3004-3010.
59. Rondou P, Haegeman G, Vanhoenacker P et al. BTB Protein KLHL12 targets the dopamine D4 receptor for ubiquitination by a Cul3-based E3 ligase. J Biol Chem 2008; 283:11083-11096.
60. Mesrobian CM, Bentley CA, Perdue SA et al. The Cul3/KLHL5 E3 Ligase regulates P60/Katanin and is required for normal mitosis in mammalian cells. Submitted.
61. Dequen F, Bomont P, Gowing G et al. Modest loss of peripheral axons, muscle atrophy and formation of brain inclusions in mice with targeted deletion of gigaxonin exon 1. J Neurochem 2008; 107:253-264.
62. Greenberg CC, Connelly PS, Daniels MP et al. Krp1 (Sarcosin) promotes lateral fusion of myofibril assembly intermediates in cultured mouse cardiomyocytes. Exp Cell Res 2008; 314:1177-1191.

Concluding Remarks

KARs function with dual identity. As ligand-gated ion channels they contribute to the postsynaptic component of the glutamatergic EPSC by depolarizing neurons. In an alternative context, KARs evince a metabotropic mode of action which does not appear to depend on their ionotropic activity *per se*. The bi-modality is a feature also seen in another iGluRs, viz. the AMPA receptor. The metabotropic operation of KARs manifests in several different ways: (i) the activation of $G_{i/o}$ protein leading to PKC stimulation; (ii) activation of a G-protein leading to stimulation of AC/cAMP/PKA signalling; (iii) activation of a G-protein leading to membrane delimited regulation (no protein kinase involvement) and (iv) G-protein-independent activation of PKA. The role of PKC and PKA in the metabotropic activity of KARs is evident in the facilitatory and inhibitory actions of KARs at central synapses with the presynaptic modulation of glutamate or GABA release and, at the same time, control of general neuronal excitability by presynaptic or postsynaptic mechanisms. Notwithstanding, the burgeoning literature on metabotropic effectors being activated by KARs working in a non-ionotropic mode, the actual mechanism of metabotropic coupling between KAR and G-protein (directly or indirectly) remaining elusive. Together with this intriguing conundrum, other key issues arising from the debate are: What are the physiological scenarios which warrant fast (ionotropic) versus slow (metabotropic) KAR action. Are there KAR subunit composition rules which determine the mode of KAR operation and receptor affinity given the differential "functional compartmentalization" of these intriguing receptors?

Antonio Rodríguez-Moreno, PhD
Talvinder S. Sihra, PhD

Kainate Receptors: Novel Signaling Insights, edited by Antonio Rodríguez-Moreno and Talvinder S. Sihra. ©2011 Landes Bioscience and Springer Science+Business Media.

Index

A

Actinfilin 115, 117-121
Active dimer 93, 96-98, 102
Afterhyperpolarization (sAHP) 32, 49, 50, 53, 54, 56
Alternative splicing 12, 116
Ambient glutamate 17, 22, 86, 106
AMPA-receptor 1, 2, 4, 5, 12, 16, 17, 27, 39, 49, 50, 59, 70, 75, 82, 94, 96, 97, 99, 101, 102, 104, 106, 108, 127
Analytical ultracentrifugation (AUC) 102, 103

B

BTB-Kelch 115, 116, 119

C

CA1 1-8, 13, 14, 39-41, 44, 46, 49-51, 53, 54, 56, 57, 60-63, 83, 84, 86, 87, 88, 116
CA3 7, 11-17, 21, 22, 39, 40-47, 49, 59-65, 82, 83, 85, 86, 88, 115
Calcineurin 74, 75
Calcium 13-15, 19, 21, 22, 73, 83, 103, 105, 106, 119
Calcium channel 13, 19, 21, 73, 105
Cell excitability 13, 27
C-fiber 69-71, 75
CN2180 120, 121
Co-immunoprecipitation 117
Cul3 115-122

D

Desensitization 70, 96, 97, 99, 101-104, 106-109
Development 2, 7, 11, 12, 14, 15, 21, 23, 27, 36, 40, 46, 49, 64, 70-72, 76, 81, 83, 84, 86, 87, 105, 115, 119, 122
Dorsal horn 69, 71-73, 75
Dorsal root ganglia (DRG) 2, 69, 70-76

E

E3 ligase 115, 116, 119, 121, 122
Epilepsy 39, 115
Excitability 2, 7, 13, 15, 19-21, 27, 35, 40, 44, 49, 50, 56, 59-64, 72, 81-83, 85, 86, 88, 115, 127

F

Fast perfusion 99

G

Gamma-aminobutyric acid (GABA) 1-8, 11-17, 19, 21-23, 27, 31, 34-36, 39, 40, 42, 44, 72, 73, 82, 105, 115, 116, 127
Globus pallidus 2, 29
GluK1 receptor 17, 19, 20, 96
GluK5 11-13, 60-65, 106, 108
GluR6 7, 8, 27-29, 31, 33, 35, 36, 49, 51, 56, 57, 60, 70-76, 82, 83, 87, 115-121
Glutamate 1, 2, 4, 5, 7, 8, 11-18, 22, 27, 39-46, 49-51, 53, 54, 57, 59, 60-65, 69, 70, 72, 73, 75, 81, 83, 86, 93-109, 115-117, 119, 121, 127
Glutamate release 1, 2, 4, 7, 8, 12, 13, 39-46, 50, 54, 72, 73, 75, 83
G protein 1, 2, 6, 7, 11, 13, 14, 18, 19, 21, 27, 29-34, 36, 39-41, 43, 46, 49, 50, 53, 54, 57, 59, 60, 62, 63, 65, 69, 72-74, 83, 86
GYKI 16, 18, 19, 52, 64, 70

H

High-affinity kainate receptor 85
Hippocampal 1-3, 5-7, 12-15, 18, 23, 27, 32, 41-43, 46, 49, 50, 53, 59, 60-62, 64, 65, 72, 75, 76, 82, 84-88, 106, 115-118, 120
Hippocampus 1-4, 7, 8, 11, 12, 14, 15, 17, 27, 29, 31, 36, 39, 40, 46, 49, 50, 60, 69, 70, 73, 75, 81-86, 88

I

Immature mossy fibre 11
Intracellular store 19, 72, 74, 76
Ion binding site 94, 96, 98, 100-106
Ionotropic action 5, 8, 13, 40, 60
IPSC 3,-6, 13, 29, 30, 32-34, 51
IsAHP 13, 32, 49-57, 59-65

K

KA2 7, 12, 27-29, 33, 35, 36, 60, 70, 71, 74-76, 116, 117, 121
Kainate (KA) 1-8, 11-14, 17-22, 27, 29, 31-35, 39, 40, 42-46, 49-57, 59, 61-63, 69-77, 81-83, 85, 86, 88, 93-97, 99, 101-109, 115, 116, 120, 121
Kainate receptor 1-3, 5, 11-14, 17-22, 27, 35, 39, 40, 49-54, 56, 57, 59, 69-77, 81-83, 85, 86, 88, 93-97, 99, 101-109, 115, 121
Knockout mice 60, 62, 63, 70, 76, 82, 109, 116, 117, 119

L

Ligand-binding domain 95, 97

M

MAP kinase (MAPK) 56, 57, 63
Mayven 115, 116, 118, 119, 121
Metabotropic action 1-3, 5-9, 11, 13-17, 21, 22, 27, 33, 36, 39-42, 44-47, 49, 50, 53, 54, 56, 57, 59-63, 65, 69, 72-76, 108, 115, 127
Molecular dynamic 103, 104
Monovalent ion 93-97, 103-105, 109
Mossy fiber 11-14, 16, 39, 46, 60-62, 64, 73, 83, 108, 115, 116
Mutant 15, 61, 96, 97, 99, 102-105, 118

N

Network activity 1, 14, 15, 46, 81, 82, 84-86, 88
Neurotransmitter 1, 2, 7, 15-17, 29, 39, 50, 60, 62, 73, 108, 115, 116
Neurotransmitter release 2, 60, 73, 108, 116
NMDA-receptor 2, 3, 17, 50, 96, 104, 105, 108, 117

O

Open probability 96, 101, 105

P

Pain 69, 75, 76, 115
Paired-pulse facilitation 16, 32
Patch-clamp 50, 62, 70
Peptidomimetic drug 119
Pertussis toxin 2, 3, 6, 11, 13, 18, 39, 40, 43, 45, 53, 54, 56, 57, 72, 74
Philantotoxin 20
Postnatal development 11, 12, 14, 15, 21, 81, 83, 86, 105
Presynaptic kainate receptor 12, 17, 18, 20, 22, 73, 76, 83
Protein kinase A (PKA) 13, 32, 34, 39, 42-46, 50, 53, 56, 57, 62, 63, 127
Protein kinase C (PKC) 1-6, 13, 14, 19, 21, 31, 32, 34, 36, 44, 45, 53, 55, 56, 57, 62, 63, 69, 72-76, 83, 88, 127
Pyramidal cell 1, 2, 6, 12, 13, 15, 16, 39-42, 46, 49-51, 53, 54, 56, 57, 60, 62, 86, 115, 116

R

Release of GABA 1, 2, 6, 11, 15, 17, 72, 73

S

Second messenger 4, 50, 62, 63, 72
Seizure 15, 23, 50, 82, 120, 121
Shank 119, 120
Spike-time dependent depression 22
Spike-time dependent potentiation 11, 22
Striatum 28, 29, 31
Substantia nigra pars compacta 33
Synaptic 1, 2, 7, 8, 11-19, 21, 22, 27, 29, 30, 32, 33, 36, 39, 40, 42, 44-46, 49, 51, 53, 54, 59, 60, 62, 69, 71, 72, 75, 76, 81-84, 86-88, 94, 99, 105, 107, 115-121
Synaptic plasticity 11, 12, 21, 42, 44, 46

U

Ubiquitination 115-119, 122
Ubiquitin-proteasome pathway 116, 117

W

Western blot 65, 87

FSC
www.fsc.org
MIX
Papier | Fördert gute Waldnutzung
FSC® C083411

Kolibri 360 GmbH
Ferdinand-Jühlke-Straße 7
99095 Erfurt, Deutschland
produktsicherheit@kolibri360.de